INTERNATIONAL CENTRE FOR MECHANICAL SCIENCES

COURSES AND LECTURES - No. 300

ADVANCES IN COMPUTATIONAL NONLINEAR MECHANICS

EDITED BY

I. St. DOLTSINIS
UNIVERSITY OF STUTTGART

SPRINGER-VERLAG WIEN GMBH

Le spese di stampa di questo volume sono in parte coperte da
contributi del Consiglio Nazionale delle Ricerche.

This volume contains 59 illustrations.

Originally published by Springer Verlag Wien-New York in 1989

In order to make this volume available as economically and as rapidly as possible the authors' typescripts have been reproduced in their original forms. This method unfortunately has its typographical limitations but it is hoped that they in no way distract the reader.

ISBN 978-3-211-82113-8 **ISBN 978-3-7091-2828-2 (eBook)**
DOI 10.1007/978-3-7091-2828-2

PREFACE

The demand for computational methods in the field of nonlinear mechanics is evident. Whereas in linear problems of practical importance the use of the computer is justified by the complexity of the domain, in the nonlinear case complicated tasks may originate from the nonlinearity itself and necessitate a computational treatment. The communication of the physical phenomenon to the computer has to be considered an essential part of the numerical solution of nonlinear problems and may have a significant effect on the relevance of the results of the calculation. Due to the subtle interaction between physics and numerics, computational nonlinear mechanics still require both a clear understanding of the physical principles underlying the system under consideration and a profound knowledge of the appropriate numerical methodologies.

It is only natural that the introduction of nonlinear computer techniques evolved from a suitable extension of linear algorithms to the analysis of moderate deviations from linear behaviour. The keyword in this context was incrementation. Subsequent developments are mainly characterised by the efforts undertaken towards a compensation of linearisation errors associated with incremental procedures. The current stage of algorithmic research and development is driven by the recognition that a numerical treatment of strongly nonlinear problems in mechanics requires specific computer models and solution techniques adjusted to the system under investigation. Also, efficiency requirements demand an adequate consideration of the architecture of the available hardware. The progress made in the development of nonlinear computer methods in conjunction with concurrent hardware developments today allows the treatment of problems in mechanics of a degree of complexity which makes remarkable nonlinearities of the initial period appear negligible. An illustrative example in this context may be the early elastoplastic analysis of structures and continua compared to current numerical simulations of complex metal forming processes of industrial relevance.

In recognition of the increasing importance of the subject for engineering and research, Professor Antoni Sawczuk in his status as a Rector of CISM proposed as early as 1984 to the Institute for Computer Applications to conceive and coordinate a CISM-course on *Computational Nonlinear Mechanics*. The proposal was accepted and has been realised in summer 1987. In retrospect, I feel that for several reasons this time offered a favourable constellation for the course. In particular, computational concepts had matured to methodologies of analysis in various fields of mechanics so that a number of distinguished research Institutions in Europe known for their pioneering work in this area were prepared to contribute to the planned event. The great interest in the lectures on *Computational Nonlinear Mechanics* given at CISM in Udine, July 1987 has indicated a demand for the publication of the material of the course as a text book. For this purpose the lectures have been elaborated to form the contributions constituting the book at hand.

As the progress in computer methods in applied mechanics and engineering equally concerns the deformation of solids and the motion of fluids, both subjects have been treated in the course within the framework of the finite element formalism. In the text, the first chapter introduces nonlinear concepts for the analysis of structures and solids undergoing quasistatic deformations, while chapter 2 is addressing the numerical treatment of problems of nonlinear dynamics, and chapter 3 deals with specific tasks in earthquake dynamics. Chapter 4 presents a survey of coupled problems; a classification is given and appropriate numerical solution techniques are discussed. As an introduction to the computer-oriented consideration of fluid motion, chapter 5 develops generalised Galerkin methods for problems dominated by convective phenomena. The text is concluded by a chapter on computational solution techniques in fluid dynamics. This last chapter describes some novel iterative methods for the solution of the compressible and incompressible Navier-Stokes equations by operator splitting.

The book should be of interest to the engineering and scientific community. The professional engineer will certainly find in it material related to his particular field and may be guided towards a solution of actual problems with the aid of the computer. The computational analyst, on the other hand, may find interest in the entire text recognising common underlying principles, similarities or even differences in the methodological approach to the various physical phenomena.

In conclusion, I wish to express my gratitude to those who have contributed to the success of the CISM-course and to the realisation of the present publication.

I.St. Doltsinis

CONTENTS

CHAPTER 1

NONLINEAR CONCEPTS IN THE ANALYSIS OF SOLIDS AND STRUCTURES

I. St. Doltsinis
University of Stuttgart, Stuttgart, FRG

Abstract

The present treatise deals with computational methods for the analysis of solids and structures on the basis of the finite element approach. It comprises solution techniques for nonlinear structures and computer simulation methodologies for large deformation processes of solids.

Unsteady contact conditions and friction phenomena are accounted for, and the possibility of an adaptation of the discretisation mesh to the deformation of the solid is considered. Applications of the developed computer-based methodology are demonstrated in the field of nonlinear dynamics, metal forming processes and continuum damage mechanics.

1. Introduction

The present account indicates the current state of the art at the Institute for Computer Applications (ICA) in the field of nonlinear computational mechanics with respect to the analysis of nonlinear solids and structures. It comprises a number of recent developments in the computer simulation of the relevant physical phenomena occuring during the course of large deformation processes in solid materials, as well as the current understanding in the methodology for the computation of nonlinear structures.

Section 2 outlines the conceptual bases of a methodology for the nonlinear analysis of solids and structures in the context of the finite element approach. At this place, only quasistatic deformation processes which are governed by the equilibrium condition are considered. An iterative solution procedure is given in general terms and is discussed with respect to its convergence properties and to the main computational steps which are to be performed in the algorithm. Numerical path-following techniques are also presented in this context and are of particular importance when limiting points at the deformation path are to be overcome, either with respect to the applied forces or to the displacements of the deforming body.

Section 3 deals with the numerical treatment of nonlinear structures. Here, a detailed definition of the kinematic and static quantities in the discretised structures is given first and is followed by an analysis of the two main procedures which are fundamental to the solution algorithm. These are the specification of the stress resultants in the deformed geometry of the discretised structure, and the gradient matrix of the system. The section is concluded with some remarks concerning the treatment of rotational degrees of freedom under the occurence of large rotations.

While in section 3 strains were assumed to be small, section 4 is concerned with the subject of the large deformations of solids. Constitutive material laws appropriate for the description of large deformation processes of solids are reviewed for the isothermal case. They comprise elastoplastic behaviour and viscous response. The computational treatment of elastoplastic solids undergoing large deformations follows the lines of section 2. In the present section, particular attention is paid to solution techniques for large deformation processes of rigid-viscous solids. Also handled is the numerical approach to the phenomena upon the occurence of contact between the deforming material and a rigid boundary. The developed procedure encompasses the normal contact condition and the appearance of friction phenomena. An important task in connection with large deformation processes concerns the possible deterioration of the discretisation mesh during the course of the computation. The subject of an adaptation of the mesh to the requirements of the solution is discussed at the end of section 4.

Applications of the computational methods to the solution of a number of selected nonlinear problems are presented in section 5. Nonlinear dynamics are treated in the context of the crash analysis for a car structure. Thermomechanically coupled problems are represented by the investigation of the three-dimensional hot-forging process for a compressor blade. The computer simulation of superplastic forming processes is demonstrated by the forming of a three-dimensional industrial component out of a plane sheet. The subject of damage and failure of elastoplastic solid materials in the course of the deformation is developed, in conclusion, within the framework of continuum damage mechanics. The theory is applied to the solution of test cases and is compared with experimental results where possible.

The purpose of the present treatise is to introduce the reader into present-day understanding of nonlinear computational mechanics of solids and guide him through recent developments in this field. In this spirit any particular details have been omitted in favour of a concise presentation of the material.

2. Methodology of Analysis

2.1 Finite Element Formulation

Consider the process of quasistatic deformation of a structure or of a solid material. In this case inertia terms can be omitted in the equations of motion, so that the deformation process is governed by the condition of equilibrium. When based on the above assumption, the appropriate starting point for the development of a finite element formulation is the principle of virtual work expressed in the customary form, cf. [1],

$$\int_V \tilde{\boldsymbol{v}}^{\mathrm{t}} \boldsymbol{f} \mathrm{d}V + \int_S \tilde{\boldsymbol{v}}^{\mathrm{t}} \boldsymbol{t} \mathrm{d}S = \int_V \tilde{\boldsymbol{\delta}}^{\mathrm{t}} \boldsymbol{\sigma} \mathrm{d}V \tag{1}$$

In (1), V is to be understood as a finite volume in the deformed configuration of the body, S is the bounding surface of the element. Kinematic and static variables appear as vector arrays and are defined by their components with respect to a cartesian coordinate system $Ox_1x_2x_3$ fixed in space. Accordingly, the vector of volume forces $\boldsymbol{f}$ and the vector of surface tractions $\boldsymbol{t}$ are represented by 3×1 column matrices comprising the respective cartesian components. The tilde refers to virtual states. Thus the 3×1 column matrix $\tilde{\boldsymbol{v}}$ comprises the cartesian components of a virtual velocity vector field. The components of the associated rate of deformation collected in the 6×1 vector array $\tilde{\boldsymbol{\delta}}$, are based on the symmetric part of the velocity gradient [2]. They correspond to the Cauchy stresses in the 6×1 vector array $\boldsymbol{\sigma}$.

In the finite element methodology [1] [3], the velocity field within each finite element is approximated as

$$\boldsymbol{v} = \boldsymbol{\omega} \boldsymbol{V}_e \tag{2}$$

where the $3n \times 1$ vector $\boldsymbol{V}_e$ comprises the velocities of the n element nodal points, and the matrix $\boldsymbol{\omega}$ contains interpolation functions. The rate of deformation may be derived from (2) in the form

$$\boldsymbol{\delta} = \boldsymbol{\alpha} \boldsymbol{V}_e \tag{3}$$

The matrix $\boldsymbol{\alpha}$ represents the customary small strain operator obtained from the element kinematics in (2) at the deformed state. Use of relations (2) and (3) for the virtual kinematic fields in (1) provides the finite element expression of the equilibrium condition,

$$\boldsymbol{Q} = \int_V \boldsymbol{\omega}^{\mathrm{t}} \boldsymbol{f} \mathrm{d}V + \int_S \boldsymbol{\omega}^{\mathrm{t}} \boldsymbol{t} \mathrm{d}S = \int_V \boldsymbol{\alpha}^{\mathrm{t}} \boldsymbol{\sigma} \mathrm{d}V = \boldsymbol{P} \tag{4}$$

The integration extends now over the volume and surface of the finite element. The vector $\boldsymbol{P}$ represents a $3n \times 1$ array comprising the forces resulting at the element nodal points from stresses $\boldsymbol{\sigma}$ in the element. Analogously, $\boldsymbol{Q}$ contains the contributions of the applied forces $\boldsymbol{f}$ and $\boldsymbol{t}$.

The velocities at the N nodal points of the finite element assembly (mesh) are collected in the $3N \times 1$ vector $\boldsymbol{V}$. Therefrom, the velocities $\boldsymbol{V}_e$ at the nodal points of the individual elements are as usual extracted by the symbolic operation,

$$\{\boldsymbol{V}_e\} = \{\mathbf{a}_e\} \boldsymbol{V} = \mathbf{a} \boldsymbol{V} \tag{5}$$

the hypervector $\{\boldsymbol{V}_e\}$ containing the velocities $\boldsymbol{V}_e$ of all individual elements. Correspondingly, the operation

$$\boldsymbol{R} = \mathbf{a}^{\mathrm{t}} \{\boldsymbol{Q}\} \tag{6}$$

furnishes the resultants of the distributed forces acting in the discretised body and places them within the $3N \times 1$ load vector $\boldsymbol{R}$ on the nodal points by accumulation of the contributions from the individual elements in $\{\boldsymbol{Q}\}$. Besides, $\boldsymbol{R}$ may comprise the actions of concentrated forces at the nodal points. The resultants of the internal stresses are supplied by the instruction

$$\boldsymbol{S} = \mathbf{a}^{\mathrm{t}} \{\boldsymbol{P}\} \tag{7}$$

which is formally equivalent to (6).

Let the $3N \times 1$ vector $\boldsymbol{X}$ comprise the cartesian co-ordinates required at each of the N nodal points for the specification of the geometry of the discretised body, and t denote the time acting as a parameter in the process of loading. A functional dependence of the load vector on the time and on the deforming geometry can then be denoted by

$$\boldsymbol{R} = \boldsymbol{R}(t, \boldsymbol{X}) \tag{8}$$

The procedure for the computation of $\boldsymbol{S}$ as resulting from (7) and the stress integral in (4) is reflected in the functional relation

$$\boldsymbol{S} = \boldsymbol{S}(\boldsymbol{\sigma}, \boldsymbol{X}) \tag{9}$$

The stress $\boldsymbol{\sigma}$ depends via the relevant material law on the strains and/or on the rate of deformation when viscous effects are apparent. It may be considered as a function of the deformed geometry and of the instantaneous velocity in the body specified by the nodal values $\boldsymbol{X}$ and $\boldsymbol{V}$ respectively. For large deformations, not only the stress $\boldsymbol{\sigma}$, but also the actual geometry $\boldsymbol{X}$ enters as an unknown quantity in (9) and leads to a nonlinear expression. Additional nonlinearities occur in the presence of large strains which are nonlinear functions of the deformed geometry. Moreover, the strains and/or the rate of deformation may be related to the stresses with a nonlinear material law.

Finally, the equilibrium condition for the deformed body at time instant t can be expressed as

$$\boldsymbol{R}(t, \boldsymbol{X}) = \boldsymbol{S}(\boldsymbol{V}, \boldsymbol{X}) \tag{10}$$

where the appearance of the velocity on the right-hand side accounts for possible viscous phenomena.

2.2 Iterative Solution Procedure

An alternative form of the equilibrium condition (10) taken at time instant $t =$ const. is the so-called residual formulation.

$$\boldsymbol{F}(\boldsymbol{X}) = \boldsymbol{R}(t, \boldsymbol{X}) - \boldsymbol{S}(\boldsymbol{X}) = \boldsymbol{o} \tag{11}$$

Here, viscous phenomena have been excluded for the present purpose. The computational treatment of viscosity is to be considered in section 4. Expression (11) is convenient for the application of the iterative scheme

$$\boldsymbol{X}_{i+1} = \boldsymbol{X}_i + \boldsymbol{H}_i \boldsymbol{F}(\boldsymbol{X}_i) \tag{12}$$

to obtain the geometry $\boldsymbol{X}$ of the deformed body at time instant t.

Following the recurrence formula in (12), the ith iteration cycle starts with an approximation $\boldsymbol{X}_i$ to the solution and supplies with the aid of the iteration matrix $\boldsymbol{H}_i$ a new estimate $\boldsymbol{X}_{i+1}$. The matrix $\boldsymbol{H}$ may be considered an auxiliary one, usually chosen with respect to best numerical properties and computational convenience. It can be seen from (12) that convergence of the iteration process implies that the equilibrium condition (11) is fulfilled.

For an assessment of the convergence properties of the iteration procedure (12), we deduce for the difference between consecutive iterations the expression

$$\delta_{i+1}\boldsymbol{X} = [\boldsymbol{I} + \boldsymbol{H}\boldsymbol{G}]\delta_i\boldsymbol{X} \tag{13}$$

where $\boldsymbol{I}$ denotes the identity matrix and

$$\boldsymbol{G} = \frac{\mathrm{d}\boldsymbol{F}}{\mathrm{d}\boldsymbol{X}} = \frac{\mathrm{d}\boldsymbol{R}}{\mathrm{d}\boldsymbol{X}} - \frac{\mathrm{d}\boldsymbol{S}}{\mathrm{d}\boldsymbol{X}} \tag{14}$$

the gradient for t = const. of the residuum $[\boldsymbol{R} - \boldsymbol{S}]$ with respect to the deforming geometry $\boldsymbol{X}$. The requirement for convergence is based on the spectral norm of the magnification matrix in (13) and reads

$$\|\boldsymbol{I} + \boldsymbol{H}\boldsymbol{G}\| < 1 \tag{15}$$

Accordingly, convergence depends on the relation of the iteration matrix $\boldsymbol{H}$ to the gradient matrix $\boldsymbol{G}$ of the system.

The constituents of the gradient matrix $\boldsymbol{G}$ in (14) may be identified as the load correction matrix

$$\boldsymbol{K}_L = \frac{\mathrm{d}\boldsymbol{R}}{\mathrm{d}\boldsymbol{X}} \tag{16}$$

and the tangential stiffness matrix of the body

$$\boldsymbol{K}_T = \frac{\mathrm{d}\boldsymbol{S}}{\mathrm{d}\boldsymbol{X}} \tag{17}$$

The matrix $\boldsymbol{K}_L$ reflects a dependence of the load vector on the deforming geometry, cf [4]. Concerning $\boldsymbol{K}_T$ one may formally deduce from (9) the expression

$$\frac{\mathrm{d}\boldsymbol{S}}{\mathrm{d}\boldsymbol{X}} = \left[\frac{\partial \boldsymbol{S}}{\partial \boldsymbol{\sigma}}\right]_X \frac{\mathrm{d}\boldsymbol{\sigma}}{\mathrm{d}\boldsymbol{X}} + \left[\frac{\partial \boldsymbol{S}}{\partial \boldsymbol{X}}\right]_\sigma \tag{18}$$

where the indices denote variables which are kept constant at the differentiation process. For an interpretation of the distinct differential expressions, $\boldsymbol{\sigma}$ is considered a stress adhering to the material. The first term in (18)

$$\boldsymbol{K}_M = \left[\frac{\partial \boldsymbol{S}}{\partial \boldsymbol{\sigma}}\right]_X \frac{\mathrm{d}\boldsymbol{\sigma}}{\mathrm{d}\mathbf{X}} \tag{19}$$

reproduces changes of the stress resultants $\boldsymbol{S}$ due to changes in the material stress during deformation. It effectively reflects the stress–strain relation of the material. The second term is the matrix

$$\boldsymbol{K}_G = \left[\frac{\partial \boldsymbol{S}}{\partial \boldsymbol{X}}\right]_\sigma \tag{20}$$

which describes changes of the stress resultants $\boldsymbol{S}$ due to changes of the geometry of the body at constant stress in the material. It is known as the geometric stiffness matrix. With the above definitions the tangential stiffness matrix (17) reads

$$\boldsymbol{K}_T = \boldsymbol{K}_M + \boldsymbol{K}_G \tag{21}$$

and the system gradient matrix (14) may ultimately be written as

$$\boldsymbol{G} = \boldsymbol{K}_L - \boldsymbol{K}_T = \boldsymbol{K}_L - [\boldsymbol{K}_M + \boldsymbol{K}_G] \tag{22}$$

When the gradient matrix $\boldsymbol{G}$ is known, the familiar Newton iterative scheme may be applied to the solution of (11) and reads

$$\boldsymbol{X}_{i+1} = \boldsymbol{X}_i - \boldsymbol{G}_i^{-1}\boldsymbol{F}(\boldsymbol{X}_i) \tag{23}$$

The Newton scheme may be considered as a particular form of the iterative procedure in (12) with

$$\boldsymbol{H}_i = -\boldsymbol{G}_i^{-1} \tag{24}$$

for the iteration matrix. For the above choice, the convergence criterion (15) predicts best convergence properties. The general iterative scheme of (12) permits, however, other convenient choices for $\boldsymbol{H}$ which may be based on approximations to the system gradient matrix $\boldsymbol{G}$.

As a matter of fact, the iterative solution algorithm based on (12) requires the implementation of two main procedures. One procedure concerns the correct statement of the equilibrium condition (11) which governs the deformed configuration. This task involves the specification of the load vector $\boldsymbol{R}$ and of the stress resultants $\boldsymbol{S}$ as a function of the deformed geometry. They are required for the evaluation of the residual vector $\boldsymbol{F}$ during iteration. The second procedure deals with the formation of the iteration matrix $\boldsymbol{H}$ from an appropriate substitute of the gradient matrix $\boldsymbol{G}$ of the system.

2.3 Numerical Path-Following Techniques

The described solution procedure should not be necessarily associated with an application of the loads appertaining to the considered time instant at once. Several reasons such as the interest in the history of deformation, strong nonlinearities, path dependence of the process call for the use of an incremental continuation technique in the actual calculation. To this end consider (11) in the following for a varying process time. In that case we may write

$$\boldsymbol{F}(\boldsymbol{X}, t) = \boldsymbol{R}(t) - \boldsymbol{S}(\boldsymbol{X}) = \boldsymbol{o} \tag{25}$$

where, for convenience, the applied loads are assumed to depend only on time. The continuation procedure starts an incremental step at the known equilibrium state 'a' of the solid – this may initially be the unloaded state – and solves (25) for the next state 'b' characterised by the incrementally advanced load parameter

$$t = {}^a t + \tau = {}^b t \tag{26}$$

associated with the load vector,

$$\boldsymbol{R} = {}^a\boldsymbol{R} + \Delta\boldsymbol{R} = {}^b\boldsymbol{R} \tag{26a}$$

and the new geometry

$$\boldsymbol{X} = {}^a\boldsymbol{X} + \Delta\boldsymbol{X} = {}^b\boldsymbol{X} \tag{27}$$

A standard solution procedure for this incremental transition follows the iteration scheme of (12) while the value of the load parameter remains fixed at $t = {}^b t$, Fig. 1. The iteration process usually starts with $\boldsymbol{X}_0 = {}^a\boldsymbol{X}$ as an estimate to the new equilibrium configuration. Subsequent iterations improve this approximation up to the desired accuracy.

It is known that nonlinear solids and in particular structures undergoing large deflections may be sensitive with respect to the application of the external loads. In this connection, a temporary change in the calculation from load control to displacement control, whenever possible, is often helpful [5]. In more general situations, the use of a variable load parameter within the incremental step appears to be advantageous [6], cf. also [7]. Thereby, the load parameter $t = {}^b t$ is adjusted to the instantaneous geometry $\boldsymbol{X} = {}^b\boldsymbol{X}$ of the deforming body in accordance with a scalar constraint equation of the form

$$f(t, \boldsymbol{X}) = 0 \tag{28}$$

which supplements the equilibrium condition (25). A general procedure for this will be outlined in what follows.

When dealing with a variable load parameter within the incremental step defined by (26) and (27), both $\boldsymbol{X}$ and t appear as unknowns in (25). Therefore, this equation is to be solved iteratively for $\boldsymbol{X}$ in conjunction with (28) constraining t, Fig. 2. Let $\boldsymbol{X}_i, t_i$ denote the result of the ith iteration cycle not necessarily satisfying the system of equations (25), (28). Linearisation of (25) at this stage yields

$$\boldsymbol{F}_i + \left[\frac{\partial \boldsymbol{F}}{\partial t}\right]_i \delta t + \left[\frac{\partial \boldsymbol{F}}{\partial \boldsymbol{X}}\right]_i \delta \boldsymbol{X} = \boldsymbol{o} \tag{29}$$

and the next iterates are

$$\boldsymbol{X}_{i+1} = \boldsymbol{X}_i + \delta \boldsymbol{X} \quad , \quad t_{i+1} = t_i + \delta t \tag{30}$$

The differential quotients in (29) are identified as

$$\frac{\partial \boldsymbol{F}}{\partial t} = \frac{\mathrm{d}\boldsymbol{R}}{\mathrm{d}t} = \boldsymbol{R}_t \tag{31}$$

which is the rate of the applied loads $\boldsymbol{R}$ with respect to the parameter t , and

$$\frac{\partial \boldsymbol{F}}{\partial \boldsymbol{X}} = -\frac{\mathrm{d}\boldsymbol{S}}{\mathrm{d}\boldsymbol{X}} = -\boldsymbol{K}_T \tag{32}$$

which in the present case is the tangential stiffness matrix of the body.

From (29),

$$\delta \boldsymbol{X} = \boldsymbol{K}_T^{-1} \left[\boldsymbol{F} + \boldsymbol{R}_t \, \delta t\right] = \delta_F \boldsymbol{X} + \delta_R \boldsymbol{X} \tag{33}$$

where, for simplicity, the subscript i is omitted here and in the following. With reference to (33), the evaluation of changes in geometry between the consecutive iterations may be composed of two partial solutions:

The contribution

$$\delta_F \boldsymbol{X} = \boldsymbol{K}_T^{-1} \boldsymbol{F} \tag{34}$$

appertains to the standard solution for a constant load fixed by $t = t_i$. The contribution

$$\delta_R \boldsymbol{X} = \left[\boldsymbol{K}_T^{-1} \boldsymbol{R}_t\right] \delta t = \boldsymbol{X}_R \, \delta t \tag{35}$$

accounts for a variation in load during iteration within the incremental step. The associated parameter variation δt can be computed from the scalar constraint equation (28) written as

$$f\,(t_i + \delta t, \boldsymbol{X}_i + \delta \boldsymbol{X}) = 0 \tag{36}$$

and substituting for $\delta \boldsymbol{X}$ expression (33) with (34) and (35). Alternatively, a linearisation of (36) by

$$f + \frac{\partial f}{\partial t}\delta t + \frac{\partial f}{\partial \boldsymbol{X}}\delta \boldsymbol{X} = 0 \tag{37}$$

furnishes

$$\delta t = -\left[f + \frac{\partial f}{\partial \boldsymbol{X}}\delta_F \boldsymbol{X}\right]\left[\frac{\partial f}{\partial t} + \frac{\partial f}{\partial \boldsymbol{X}}\boldsymbol{X}_R\right]^{-1} \tag{38}$$

The computation of δt completes the iteration cycle. The new values of the iteration variables are given by (30). If they do not satisfy the system of the governing equations (25) and (28), a next iteration cycle is started.

It should be noticed that the outlined methodology is not restricted to a linear dependence of the applied loads $\boldsymbol{R}$ on the parameter t. The customary linear case is characterised by

$$\boldsymbol{R} = \boldsymbol{R}_0 t \quad , \quad \boldsymbol{R}_t = \boldsymbol{R}_0 = \text{const.} \tag{39}$$

The present treatment may be extended to loads which depend also on the deforming geometry. Under these conditions, $\boldsymbol{R}_t$ in (31) is to be understood as a partial differential, and the stiffness matrix $\boldsymbol{K}_T$ in (32) has to be supplemented by the load correction matrix $\boldsymbol{K}_L$ of (16) in accordance with (22) for the system's gradient matrix $\boldsymbol{G}$ defined in (14).

For an application of the method the constraint equation (28) must be specified. To this end consider first the adjustment of the externally applied loading to a displacement prescribed in a single degree of freedom in the structure [8]. In this case the constraint equation (28) assumes the form

$$(x - {}^a x) - s = 0 \tag{40}$$

where s denotes the prescribed increment in x. The calculation procedure starts using the state at the beginning of the increment as an estimate for the solution: $\boldsymbol{X}_0 = {}^a\boldsymbol{X}, t_0 = {}^a t$. When $\boldsymbol{F}(\boldsymbol{X}_0, t_0)$ is not exactly zero, both $\delta_F \boldsymbol{X}$ and $\boldsymbol{X}_R$ are obtained by (34) and (35) respectively and contribute to $\delta \boldsymbol{X}$ in (33). Let $x_0 + \delta x$

be the value thus obtained for x in (40). Then, detailing δx according to the second expression in (33) and observing (35), we deduce for the load parameter

$$\delta t = t_1 - t_0 = -\frac{\delta_F x - s}{x_R} \tag{41}$$

The described first iteration cycle is often called the predictor step in the literature. Corrections obtained by subsequent iterations ($i > 0$) yield

$$\delta t = t_{i+1} - t_i = -\frac{\delta_F x}{x_R} \tag{42}$$

It follows, as a consequence of (41) and (42), that the entire increment s of the prescribed displacement is applied in the initial iteration cycle of the procedure.

An alternative type of constraint for the incremental changes in geometry is specified by the equation,

$$f = [\boldsymbol{X} - {}^a\boldsymbol{X}]^t \, [\boldsymbol{X} - {}^a\boldsymbol{X}] - s^2 = 0 \tag{43}$$

cf. [9]. In contrast to (40) the above constraint involves the entire vector of the incremental displacements. By the observance of (43) the applied forces are adjusted to a prescribed value s for the length of the incremental displacement vector of the system.

The iteration process is started as previously with $\boldsymbol{X}_0 = {}^a\boldsymbol{X}$, $t_0 = {}^a t$, Fig. 3a. The result of the first iteration, ${}^a\boldsymbol{X} + \delta\boldsymbol{X}$, is substituted for $\boldsymbol{X}$ in (43) which with reference to (33) becomes a quadratic equation for the associated $\delta t = t_1 - t_0$ and reads,

$$\boldsymbol{X}_R^t \boldsymbol{X}_R (t_1 - t_0)^2 + 2\delta_F \boldsymbol{X}^t \boldsymbol{X}_R (t_1 - t_0) + \delta_F \boldsymbol{X}^t \delta_F \boldsymbol{X} - s^2 = 0 \tag{44}$$

As the values $\boldsymbol{X}_0 = {}^a\boldsymbol{X}$, $t_0 = {}^a t$ mark an equilibrium state, $\boldsymbol{F}_0$ and the associated $\delta_F \boldsymbol{X}$ theoretically vanish so that (44) may be simplified, yielding

$$\delta t = t_1 - t_0 = \pm \frac{s}{\left(\boldsymbol{X}_R^t \boldsymbol{X}_R\right)^{1/2}} \tag{45}$$

The sign of expression (45) defines the direction of the incremental transition.

For iterations $i > 0$ the loading parameter t may be determined by solving either the quadratic equation (44) or the linear form derived from it by means of (37). In the latter case specification of (38) for (43) furnishes

$$\delta t = t_{i+1} - t_i = -\frac{\frac{f}{2} + [\boldsymbol{X}_i - {}^a\boldsymbol{X}]^t \delta_F \boldsymbol{X}}{[\boldsymbol{X}_i - {}^a\boldsymbol{X}]^t \boldsymbol{X}_R} \tag{46}$$

Instead of observing the constraint equation (43) as by (46), one may apply alternative corrections which enforce changes of the displacement vector along a certain prescribed direction. As an example, the direction for the changes $[\boldsymbol{X}_{i+1} - \boldsymbol{X}_i]$ may be chosen to be perpendicular to the prediction $[\boldsymbol{X}_1 - \boldsymbol{X}_0]$ for the displacement increment, Fig. 3b. The expression of this condition is, cf. [6]

$$[\boldsymbol{X}_1 - \boldsymbol{X}_0]^t \, [\boldsymbol{X}_{i+1} - \boldsymbol{X}_i] = 0 \tag{47}$$

from which the variations of the loading parameter are deduced as

$$\delta t = t_{i+1} - t_i = -\frac{[\boldsymbol{X}_1 - \boldsymbol{X}_0]^t \delta_F \boldsymbol{X}}{[\boldsymbol{X}_1 - \boldsymbol{X}_0]^t \boldsymbol{X}_R} \tag{48}$$

An obvious modification to the above correction procedure illustrated in Fig. 3c, is obtained by replacing in (47) and (48) $\boldsymbol{X}_1$ by $\boldsymbol{X}_i$, the latest available iterate of $\boldsymbol{X}$, cf. [10].

The implementation of (48) requires storage of previous solutions for $\boldsymbol{X}$; the same applies to the mentioned modification. In [11] a variant of the constraint condition has been proposed which, in the corrector stage, bases the direction for $\delta\boldsymbol{X} = \boldsymbol{X}_{i+1} - \boldsymbol{X}_i$ solely on data from the actual equilibrium iteration for the system. For this purpose changes of the displacements during iteration are constrained in what follows by

$$[\delta_R \boldsymbol{X}]^t \delta \boldsymbol{X} = 0 \tag{49}$$

This condition expresses normality between $\delta\boldsymbol{X}$ and $\delta_R\boldsymbol{X}$, the displacements caused by variations δt of the loading parameter. By virtue of (35) $\delta_R\boldsymbol{X}$ is equal to $\boldsymbol{X}_R\delta t$ so that the aforementioned normality transfers to $\boldsymbol{X}_R$, Fig. 3d. With reference to (33) $\delta\boldsymbol{X}$ in (49) is composed of the two partial solutions $\delta_F\boldsymbol{X}$ and $\delta_R\boldsymbol{X} = \boldsymbol{X}_R\delta t$. After the respective substitution, (49) can be solved for the variation of the loading parameter and yields

$$\delta t = t_{i+1} - t_i = -\frac{\boldsymbol{X}^t{}_R \delta_F \boldsymbol{X}}{\boldsymbol{X}^t{}_R \boldsymbol{X}_R} \tag{50}$$

As a matter of fact, only $\delta_F\boldsymbol{X}$ and $\boldsymbol{X}_R$, which are quantities computed in the current iteration, appear in expression (50).

3. Nonlinear Analysis of Structures

3.1 Kinematic and Static Quantities

Consider a structure discretised by finite elements and let the vector

$$\boldsymbol{X} = \{\boldsymbol{x}_1 \ \boldsymbol{x}_2 \ \dots \ \boldsymbol{x}_i \ \dots \ \boldsymbol{x}_N\} \tag{51}$$

comprise the cartesian coordinates $\boldsymbol{x}_i$ of the N nodal points with respect to the basic system $Ox_1x_2x_3$. If the initial geometry of the structure is specified by the vector ${}^0\boldsymbol{X}$, then the deformed geometry is given by

$$\boldsymbol{X} = {}^0\boldsymbol{X} + \boldsymbol{r} \tag{52}$$

where

$$\boldsymbol{r} = \{\boldsymbol{u}_1 \ \boldsymbol{u}_2 \ \dots \ \boldsymbol{u}_i \ \dots \ \boldsymbol{u}_N\} \tag{53}$$

is the vector of nodal displacements of the structure. From the vector array $\boldsymbol{r}$ in (53) the displacements $\boldsymbol{u}_k$ of the n nodal points of a single element are extracted and form the vector

$$\boldsymbol{\varrho} = \{\boldsymbol{u}_1 \ \boldsymbol{u}_2 \ \dots \ \boldsymbol{u}_k \ \dots \ \boldsymbol{u}_n\} = \mathbf{a}_e\boldsymbol{r} \tag{54}$$

Analogously, the coordinates $\boldsymbol{x}_k$ of the element nodal points are obtained from the vector $\boldsymbol{X}$ in (51). (The notation $\boldsymbol{r} = \boldsymbol{U}, \boldsymbol{\rho} = \boldsymbol{U}_e = \mathbf{a}_e\boldsymbol{U}$ is traditional [1].)

A classification of nonlinear (finite) kinematics may be achieved by the distinction between deformation and rigid motion of the element. This is in general not a simple task and cannot be treated uniquely. Our present considerations are restricted to an elementary discussion of the algorithmic concepts under the aspect of nonlinear kinematics. To this end we introduce in the following the motion of a local reference system $(O'x_1'x_2'x_3')_e$ which follows a rigid translation and rotation of the individual element [5]. In addition to the basic coordinates $\boldsymbol{x}_k$ of the element nodal points, local coordinates $\boldsymbol{x}_k'$ are defined with respect to this system. Local and basic coordinates of element nodal points may be related by

$$\boldsymbol{x}_k' = \boldsymbol{T}\,[\boldsymbol{x}_k - \boldsymbol{x}_0] \tag{55}$$

at any stage of the deformation process. The vector $\boldsymbol{x}_0$ specifies the position of the origin of the local system for the element at the current state and the orthogonal matrix $\boldsymbol{T}$ its orientation. The local displacements $\boldsymbol{u}_k'$ of the nodal points of the element are collected in the vector array

$$\varrho' = \{u'_1 \; u'_2 \; \dots \; u'_k \; \dots \; u'_n\} \tag{56}$$

and are defined as

$$\begin{aligned} u'_k &= x'_k - {}^0x'_k = \\ &= T\,[u_k - u_0] + \left[T - {}^0T\right]\left[{}^0x_k - {}^0x_0\right] \end{aligned} \tag{57}$$

where the quantities ${}^0x'_k$ and ${}^0x_0$, ${}^0T$ refer to the initial state of the element. In the case of homogeneously deforming finite elements, rigid motion may be entirely eliminated in the local displacements defined by (57). Under more general conditions, the movement of the local reference system follows as closely as possible the rigid displacements of the element. By virtue of their definition, the displacements in (56) do not involve large rigid body motion. They contribute mainly to the strains

$$\gamma' = \gamma(\varrho') \tag{58}$$

which are arranged as a vector array. For infinitesimal strains, equation (58) expresses a linear relation between local displacements ϱ' and strains γ'. When strains are large, this relation must be nonlinear in conformity with suitable deformations of finite strain, cf. [2,20]. The strains in the element must ultimately be derived from the global displacements ϱ, and this leads in any case to a nonlinear dependence whenever the orientation matrix T has to be considered a function of the displacement in the transition from the fixed basic system to the moving local one by (57).

A linear relation can be stated only between the current rate of deformation in the vector δ' (velocity strain) in the element and the velocities $\dot{\varrho}$ of its nodes. One may write

$$\delta' = \alpha'\,\dot{\varrho}' = \alpha'\,a\,\dot{\varrho} \tag{59}$$

where α' denotes the customary small strain transformation matrix in the local system obtained from the element kinematics at the deformed state. The relation

$$\dot{\varrho}' = a\,\dot{\varrho} \tag{60}$$

reflects the time rate of (57) or (55)

$$\dot{u}'_k = T\,[\dot{u}_k - \dot{u}_0] + \dot{T}\,[x_k - x_0] \tag{61}$$

Here $\dot{\boldsymbol{u}}_0, \dot{\boldsymbol{T}}$ define the instantaneous velocity of the local element system and are connected with the rigid body part of the basic element velocites $\dot{\boldsymbol{\varrho}}$.

The rate of deformation $\boldsymbol{\delta}'$ which derives from the element velocities via (59) corresponds to a Cauchy stress $\boldsymbol{\sigma}'$ by means of the virtual work principle (1). The stress $\boldsymbol{\sigma}'$ defined in the local system of the element is transformed to resultant forces $\boldsymbol{P}$ at the element nodal points with components in the basic structural system of reference by

$$\boldsymbol{P} = \boldsymbol{a}^{\mathrm{t}} \boldsymbol{P}' = \boldsymbol{a}^{\mathrm{t}} \int_V \boldsymbol{\alpha}'^{\mathrm{t}} \boldsymbol{\sigma}' \mathrm{d}V \tag{62}$$

Operation (62) involves the transformation from stresses to nodal forces of (4) as applied to the local system in conformity with the kinematics of (51). The subsequent transition to the basic system is dual to the kinematic relation (60).

The vectors $\boldsymbol{P}, \boldsymbol{P}'$ are defined in analogy to $\boldsymbol{\varrho}, \boldsymbol{\varrho}'$ in (54) and (56) respectively. The forces at the N nodal points of the discretised structure resulting from stresses are obtained by accumulation of contributions from the individual elements as by (7).

3.2 Specification of the Stress Resultants in the Deformed Geometry

The most important step in the formulation of geometrically nonlinear problems is the correct statement of the equilibrium condition (10) which governs the deformed configuration of the structure. This task concerns the specification of the applied loads $\boldsymbol{R}$ and of the stress resultants $\boldsymbol{S}$ as a function of the deformed geometry. A detailed consideration of the external loads requires information about the particular dependence on the geometry and will not be exercised here. The stress resultants are given by (7) and (62) as a function of the stress $\boldsymbol{\sigma}'$ within the individual elements and of the geometry. For an expression exclusively in terms of the deformed geometry, a suitable material law must be assumed with reference to the local element system. The present considerations concern slender structures which may undergo large deflections under the applied loads while strains remain small. If the material behaviour is elastic, the stress $\boldsymbol{\sigma}'$ is a unique function of the elastic strain $\boldsymbol{\epsilon}'$,

$$\boldsymbol{\sigma}' = \boldsymbol{\sigma}(\boldsymbol{\epsilon}') \tag{63}$$

The small strain in the element is obtained from the local displacements of its nodes by application of the first linear relation in (59). One has

$$\boldsymbol{\epsilon}' = \boldsymbol{\alpha}' \boldsymbol{\varrho}' = \boldsymbol{\alpha}' \boldsymbol{\varrho}'(\boldsymbol{\varrho}) \tag{64}$$

the operator $\boldsymbol{\alpha}'$ being taken either at the initial or at the deformed element geometry. Large rotations affect in (64) the determination of the local displacements

$\boldsymbol{\varrho}'$ from the basic ones $\boldsymbol{\varrho}$ by means of the nonlinear relation (57). With (63) and (64) in conjunction with (52) the required specification of the stress resultants $\boldsymbol{S}$, (9), in terms of $\boldsymbol{X}$ is completed. Thus, equations (10), (11) are established and may be solved for the geometry of the deformed structure once that the external loads are given. Should one be interested in the history of deformation, an incremental computation is appropriate, cf. section 2.3. Then starting with a known stage 'a' at the beginning of the increment, the subsequent stage 'b' at the end of the increment is obtained by increasing the external loads in accordance with (26) and solving (11) for the new equilibrium geometry of (27). Such an incrementation may be advantageous from the computational point of view, although it is not imperative in the present elastic case.

If the material obeys a constitutive relation of the rate-type

$$\dot{\boldsymbol{\sigma}}' = \boldsymbol{\kappa}'\boldsymbol{\delta}' \tag{65}$$

which is non-integrable due to a possible path-dependence of the deformation process (e.g. elastoplasticity), a trace of the loading history via incrementation is mandatory for the computational process. In view of the finite increments in the calculation, the constitutive equation (65) requires an integration with respect to time according to

$$\Delta\boldsymbol{\sigma}' = \int_a^b \boldsymbol{\kappa}'\boldsymbol{\delta}'\mathrm{d}t = {}^{\zeta}\boldsymbol{\kappa}'\Delta\boldsymbol{\gamma}' \tag{66}$$

Here ${}^{\zeta}\boldsymbol{\kappa}'$ denotes the material stiffness matrix evaluated at a state within the increment defined by the value of the parameter,

$$0 \leq \zeta \leq 1 \tag{67}$$

and

$$\Delta\boldsymbol{\gamma}' = \int_a^b \boldsymbol{\delta}'\mathrm{d}t \tag{68}$$

is the strain increment. Integration of (59) may be performed as by,

$$\Delta\boldsymbol{\gamma}' = \boldsymbol{\alpha}'\Delta\boldsymbol{\varrho}' = {}^{\zeta}[\boldsymbol{\alpha}'\boldsymbol{a}]\,\Delta\boldsymbol{\varrho} \tag{69}$$

which provides a convenient extension of the customary small deformation expression to finite increments. According to the small strain assumption, the operator $\boldsymbol{\alpha}'$ in (69) might be referred to the initial element geometry. It is however computationally preferable to perform the entire strain operation at a unique state. This provides an approximation also to large strain situations. In contrast to (69), an exact relation for the strain increment can be based on the nonlinear equation (57), but is inconvenient.

In the elastoplastic case the rate of deformation consists of an elastic part and an inelastic part. Possible discontinuities in the material stiffness $\boldsymbol{\kappa}'$ require a distinction between the elastic and the plastic deformation process in the integral in (66). The incremental expression of this integral may also stand for alternative operations leading to the same essential result [22].

The stress resultants $\boldsymbol{S}$ at the end of the loading step as represented by (9) are obtained with the accumulated stresses

$$\boldsymbol{\sigma}' = {}^a\boldsymbol{\sigma}' + \Delta\boldsymbol{\sigma}' = {}^b\boldsymbol{\sigma}' \tag{70}$$

which are expressed in terms of the deformed geometry by the constitutive equation (66).

Concluding the present discussion, we remark that the above incremental procedure can be adapted also to elastic material behaviour. Application of (66) and (69) permits a simple extension of linear software to geometrically nonlinear problems [14]. The incremental relations (66), (69) are evaluated with the geometry at the middle of the increment, $\zeta = 1/2$. Due to the path-independence of elastic deformation processes, these relations may be alternatively associated with the beginning and the end of an iteration cycle within the loading step.

3.3 Remarks on the Gradient Matrix of the System

The tangential stiffness matrix $\boldsymbol{K}_T$, (17), of the discretised structure is composed of contributions from individual elements as also the stress resultants $\boldsymbol{S}$ are in (7). Formally,

$$\boldsymbol{K}_T = \{\mathbf{a}_j\}^t \left\lceil \frac{\mathrm{d}\boldsymbol{P}_j}{\mathrm{d}\boldsymbol{\varrho}_j} \right\rfloor \{\mathbf{a}_j\} = \{\mathbf{a}_j\}^t \lceil \boldsymbol{k}_{Tj} \rfloor \{\mathbf{a}_j\} \tag{71}$$

where the element stiffness matrix

$$\boldsymbol{k}_T = \left[\frac{\partial \boldsymbol{P}}{\partial \boldsymbol{\sigma}'}\right]_{\varrho} \frac{\mathrm{d}\boldsymbol{\sigma}'}{\mathrm{d}\boldsymbol{\varrho}} + \left[\frac{\partial \boldsymbol{P}}{\partial \boldsymbol{\varrho}}\right]_{\sigma'} = \boldsymbol{k}_M + \boldsymbol{k}_G \tag{72}$$

conforms with (18), (21) and elucidates the meaning of the respective expressions at the structural level. The first term in (72) is the element contribution to the structural matrix $\boldsymbol{K}_M$, (19). Considering (62), (60) it may be formed as

$$\boldsymbol{k}_M = \boldsymbol{a}^{\mathrm{t}} \left[\frac{\partial \boldsymbol{P}'}{\partial \boldsymbol{\sigma}'}\right]_{\varrho'} \frac{\mathrm{d}\boldsymbol{\sigma}'}{\mathrm{d}\boldsymbol{\varrho}'} \boldsymbol{a} = \boldsymbol{a}^{\mathrm{t}} \boldsymbol{k}'_M \boldsymbol{a} \tag{73}$$

The local matrix $\boldsymbol{k}'_M$ represents the straining stiffness of the element. It is obtained from (62) and reads

$$\boldsymbol{k}'_M = \int_V \boldsymbol{\alpha}'^{\mathrm{t}} \frac{\mathrm{d}\boldsymbol{\sigma}'}{\mathrm{d}\boldsymbol{\gamma}'} \boldsymbol{\alpha}' \mathrm{d}V \tag{74}$$

An evaluation of (74) requires the functional dependence of the stress $\boldsymbol{\sigma}'$ on the strain $\boldsymbol{\gamma}'$.

The second term in (72) is the element contribution to the geometric stiffness matrix $\boldsymbol{K}_G$, (20), of the structure. With reference to (62) we may write

$$\boldsymbol{k}_G = \left[\frac{\partial \boldsymbol{P}}{\partial \boldsymbol{\varrho}}\right]_{\sigma'} = \left[\frac{\partial}{\partial \boldsymbol{\varrho}} \left(\boldsymbol{a}^{\mathrm{t}} \boldsymbol{P}'\right)\right]_{P'} + \boldsymbol{a}^{\mathrm{t}} \left[\frac{\partial \boldsymbol{P}'}{\partial \boldsymbol{\varrho}'}\right]_{\sigma'} \boldsymbol{a} \tag{75}$$

Variations in the basic components of the element stress resultants as a result of rigid rotations are defined by the first constituent in (75). The effect of variations in the local element geometry is reflected in the second constituent. For the assumed small strains this effect should vanish when the forces $\boldsymbol{P}'$ in (62) are calculated with the initial element geometry. If, however, for convenience in the computation a varying geometry is used in this operation, the second term in (75) is active even in the small strain case. Its inclusion in the stiffness matrix and respective consideration in the iteration matrix contributes to an improvement of the numerical behaviour.

3.4 Rotational Degrees of Freedom

Most of the important geometrically nonlinear applications include rotational degrees of freedom. With respect to this, the previous considerations require some modifications. Given, as usually, a finite element with infinitesimal kinematics its degrees of freedom in the basic reference system are conveniently defined in the form of velocities

$$\dot{\boldsymbol{\varrho}} = \{\dot{\boldsymbol{u}}_k \, , \, \dot{\boldsymbol{\varphi}}_k\} \tag{76}$$

and comprise translational velocities $\dot{\boldsymbol{u}}_k$ and rotational velocities $\dot{\boldsymbol{\varphi}}_k$. Accordingly, in the local element system (60) assumes the form

$$\dot{\boldsymbol{\varrho}}' = \{\dot{\boldsymbol{u}}'_k \, , \, \dot{\boldsymbol{\varphi}}'_k\} = \lceil \boldsymbol{a}_u \boldsymbol{a}_\varphi \rfloor \{\dot{\boldsymbol{u}}_k \, , \, \dot{\boldsymbol{\varphi}}_k\} \tag{77}$$

The rate of deformation (59) in the element reads then

$$\boldsymbol{\delta}' = \boldsymbol{\alpha}'_u\{\dot{\boldsymbol{u}}'_k\} + \boldsymbol{\alpha}'_\varphi\{\dot{\boldsymbol{\varphi}}'_k\} = \left[\boldsymbol{\alpha}'_u\boldsymbol{\alpha}'_\varphi\right]\{\dot{\boldsymbol{u}}'_k\,,\,\dot{\boldsymbol{\varphi}}'_k\} \tag{78}$$

A transition to finite increments of strain is performed in conformity with (69). Thus,

$$\Delta\boldsymbol{\gamma}' = {}^\zeta\left[\boldsymbol{\alpha}'_u\boldsymbol{\alpha}'_\varphi\right]\;{}^\zeta\left\lceil\boldsymbol{a}_u\boldsymbol{a}_\varphi\right\rfloor\{\Delta\boldsymbol{u}_k\,,\,\Delta\boldsymbol{\varphi}_k\} \tag{79}$$

This type of integration of (68) retains the operations inherent to the linear element kinematics. Using (79) the stress increment (66) is determined, and with (70) the stress $\boldsymbol{\sigma}'$ at the end of the incremental step. The calculation of the stress resultants at the element nodal points requires an appropriate extension of (62) to the degrees of freedom presently considered. We deduce,

$$\boldsymbol{P} = \left\lceil\boldsymbol{a}_\mu\boldsymbol{a}_\varphi\right\rfloor^{t}\boldsymbol{P}' = \left\lceil\boldsymbol{a}_u\boldsymbol{a}_\varphi\right\rfloor^{t}\int_V\left[\boldsymbol{\alpha}'_u\boldsymbol{\alpha}'_\varphi\right]^{t}\boldsymbol{\sigma}'\mathrm{d}V \tag{80}$$

As a consequence of the appearance of the accumulated stress $\boldsymbol{\sigma}'$ in the integral in (80), the element nodal forces $\boldsymbol{P}$ exhibit a direct dependence on the increment of the generalised displacements. For a distinct element the latter reads

$$\Delta\boldsymbol{\varrho} = \{\Delta\boldsymbol{u}_k,\;\Delta\boldsymbol{\varphi}_k\} \doteq \boldsymbol{a}_e\{\Delta\boldsymbol{u}_i\,,\,\Delta\boldsymbol{\varphi}_i\} \tag{81}$$

Besides, the evaluation of (80) requires the deformed geometry which must be determined from the incremental displacements of (81). If the element geometry is specified exclusively by the position of the element nodal points (e.g. facette elements [15]), then the deformed geometry of the discretised structure is given by

$$\boldsymbol{X} = \{{}^a\boldsymbol{x}_i + \Delta\boldsymbol{u}_i\} = {}^b\boldsymbol{X} \tag{82}$$

Here, only an update by the incremental translations $\Delta\boldsymbol{u}$ is needed. Following the accumulation procedure of (7), the stress resultants at the mesh nodal points obey then the functional dependence

$$\boldsymbol{S} = \boldsymbol{S}\left(\boldsymbol{\sigma}',{}^0\boldsymbol{x}_i + \boldsymbol{u}_i\right) = \boldsymbol{S}\left(\boldsymbol{\sigma}',\boldsymbol{u}_i\right) \tag{83}$$

Consequently, the rotational degrees of freedom contribute solely to the stress and to the matrix $\boldsymbol{K}_M$ of (19), but do not affect the geometric stiffness matrix $\boldsymbol{K}_G$ of (20).

A specification of the element geometry may require in addition to the position of the nodal points also the orientation of a triad of orthogonal basis vectors

$$C = [c_1\, c_2\, c_3] \tag{84}$$

at each nodal point. Then the transition from the initial state to the deformed state implies the operation

$$C = T\, {}^0C \tag{85}$$

where T denotes the rotation matrix at the element nodal point. In an incremental procedure,

$$C = \Delta T\, {}^aC = \Delta T\, {}^aT\, {}^0C = {}^bC \tag{86}$$

The update of the orientation triad presumes the formation of the incremental rotation matrix ΔT as a function of the incremental rotations $\Delta\varphi$. For specific forms of the relation

$$\Delta T = T(\Delta\varphi) \tag{87}$$

we refer, for instance, to [16].

In the present case, the rotational degrees of freedom affect both stress and geometry so that in contrast to (83) the stress resultants at the mesh nodal points exhibit the functional dependence

$$S = S\left(\sigma', {}^0x_i + u_i, T_i\, {}^0C_i\right) = S\left(\sigma', u_i, \varphi_i\right) \tag{88}$$

The explicit appearance of the rotations in (88) transfers consistently also to the formation of the geometric stiffness matrix. Nevertheless, a geometric stiffness matrix based only on the translational degrees of freedom provides a convenient approximation to the complete one, cf. [13].

4. Large Deformations of Solids

4.1 Constitutive Concepts

Consider once more the basic equation (1) which expresses the principle of virtual work for quasistatic deformation processes. In order to relate the stress on the right-hand side in (1) to the kinematic variables involved in the deformation process, the constitutive equations for the particular material must be accounted for. The present section is particular concerned with the constitutive aspect in connection with large deformations. In this context, a representative selection of constitutive material laws is discussed briefly and remarks are given with respect to computational implications. These laws are applicable to large strain elastoplastic problems and to rate-dependent phenomena under isothermal conditions. For constitutive material laws under non-isothermal situations, the reader may be referred to [29, 31].

A class of constitutive equations for large strain elastoplasticity frequently employed in numerical procedures is related to the description of elastoplastic materials proposed by Hill, [17], leading to rate equations between stress and strain. In this connection the Cauchy stress and the rate of deformation may be considered as the basic variables.

The rate of deformation in the solid is specified by the entities in the vector array

$$\boldsymbol{\delta} \;=\; \left\{\delta_{11}\;\delta_{22}\;\delta_{33}\;\sqrt{2}\delta_{12}\;\sqrt{2}\delta_{23}\;\sqrt{2}\delta_{13}\right\} \tag{89}$$

which correspond to the Cauchy stresses in the vector

$$\boldsymbol{\sigma} \;=\; \left\{\sigma_{11}\;\sigma_{22}\;\sigma_{33}\;\sqrt{2}\sigma_{12}\;\sqrt{2}\sigma_{23}\;\sqrt{2}\sigma_{13}\right\} \tag{90}$$

For our present purpose, we interpret the rate of deformation $\boldsymbol{\delta}$ as a strain rate $\dot{\boldsymbol{\gamma}}$ which may be additively composed of an elastic part $\dot{\boldsymbol{\varepsilon}}$ and an inelastic part $\dot{\boldsymbol{\eta}}$. Thus,

$$\boldsymbol{\delta} = \dot{\boldsymbol{\gamma}} = \dot{\boldsymbol{\varepsilon}} + \dot{\boldsymbol{\eta}} \tag{91}$$

A rate-elastic relation connects a material rate of the stress – e.g. the Jaumann co-rotational rate of the Cauchy stress $\breve{\boldsymbol{\sigma}}$ – with the elastic part of the strain rate, cf. [2]. For a given strain rate $\dot{\boldsymbol{\gamma}}$ and inelastic strain rate $\dot{\boldsymbol{\eta}}$ one may then write

$$\breve{\boldsymbol{\sigma}} = \boldsymbol{g}(\dot{\boldsymbol{\gamma}}, \dot{\boldsymbol{\eta}}) = \boldsymbol{\kappa}[\dot{\boldsymbol{\gamma}} - \dot{\boldsymbol{\eta}}] \tag{92}$$

The customary linear rate-elastic constitutive expression appears as a particular form of the general functional dependence in (92). The current rate-elastic stiffness matrix of the material is denoted by the symbol $\boldsymbol{\kappa}$.

In order to complete the constitutive description in (92), the inelastic strain rate $\dot{\boldsymbol{\eta}}$ must be specified. It may be of plastic, viscoplastic or of viscous origin. In the case of an inviscid plastic material, for example, a specification of $\dot{\boldsymbol{\eta}}$ relies as usual on the notion of a yield condition, a hardening rule and a flow rule. For consistency in the formulation, the Cauchy stress is used throughout in the present constitutive concept as the appropriate static variable. The stress $\boldsymbol{\sigma}$ required in (1) is to be determined by an integration of $\breve{\boldsymbol{\sigma}}$ with respect to time. The integration procedure is an approximate, incremental one. As $\breve{\boldsymbol{\sigma}}$ in the material law (92) is defined from the material point of view and follows the instantaneous spin of the velocity field in the continuum, while $\boldsymbol{\sigma}$ in the virtual work expression (1) refers to a spatially fixed frame, the integration must properly account for the finite kinematics (rotations) of the deformation process of the solid, cf. [24] for example. There are two possibilities in this connection: One may either transfer the co-rotational rate $\breve{\boldsymbol{\sigma}}$ to the ordinary rate $\dot{\boldsymbol{\sigma}}$ and perform the integration on the latter, or one may alternatively transfer $\boldsymbol{\sigma}$ to the rotated material system, perform an incremental integration of $\breve{\boldsymbol{\sigma}}$ and then transfer the result back to the spatially fixed frame.

Alternative methodologies [18, 19] establish a total-elastic constituent of the material which is characterised by an elastic relation between stress and the elastic part of strain. Elastic strains $\boldsymbol{\varepsilon}$ may be defined for given strains $\boldsymbol{\gamma}$ and inelastic strains $\boldsymbol{\eta}$. Therefore,

$$\boldsymbol{\sigma} = \boldsymbol{f}(\boldsymbol{\gamma}, \boldsymbol{\eta}) = \boldsymbol{\kappa}[\boldsymbol{\gamma} - \boldsymbol{\eta}] \tag{93}$$

where the customary linear expression represents a particular form of the general relation in (93). For economy of presentation, the same symbols have been used in (92) and (93). However, the rate variables in (92) may be defined independently of the total variables in (93).

The general functional form and the variables in (93) may be specified, so that they comply with particular constitutive assumptions. The material description according to Lee [18] takes advantage of the fact that the elastic properties of metals may be considered, to a certain extent, independent of preceding inelastic deformation. This, however, applies to strains and stresses which are referred to the so-called current stress-free configuration of the material. The stress-free configuration is based on an imaginary elastic local unloading of the deformed material and reflects the result of inelastic motion; it is generally discontinuous. The complete specification of the stress-free configuration as required by the material model is part of this constitutive concept.

Elastic and inelastic strains may be defined in accordance with the afore-mentioned partitioning of the deformation process via local unloading. The total strain

$$\boldsymbol{\gamma} = \boldsymbol{\varepsilon}_G + \boldsymbol{\eta}_A \tag{94}$$

may be conceived as a Green-type strain referring to the unloaded configuration, cf. [20, 21, 25]. This reference configuration is common to the elastic part of the strain $\boldsymbol{\varepsilon}_G$ – which turns out to be a Green strain – and to the inelastic part $\boldsymbol{\eta}_A$ which is in accordance with an Almansi definition, cf. [2]. For consistency in the formulation, we use as a static variable in the constitutive description the symmetric Piola-Kirchhoff stress referred to the stress-free configuration. This stress satisfies the virtual work expression in the form

$$^{f}\left[\frac{1}{\varrho}\overset{\circ}{\boldsymbol{\gamma}}^{t}\boldsymbol{\sigma}_{P}\right]=\frac{1}{\varrho}\boldsymbol{\delta}^{t}\boldsymbol{\sigma} \tag{95}$$

The superscript f indicates that the stress-free configurations is taken as reference, and $\overset{\circ}{\boldsymbol{\gamma}}$ denotes the time rate of the strain $\boldsymbol{\gamma}$ in (94) determined while the current stress-free configuration is assumed to be fixed.

Relation (95) transfers the Piola-Kirchhoff stress $\boldsymbol{\sigma}_P$ on the left-hand side to the Cauchy stress $\boldsymbol{\sigma}$ on the right-hand side, and vice-versa. This equation may be utilized in the integral on the right-hand side in (1), and ultimately leads to an alternative evaluation for the stress resultants $\boldsymbol{S}$ in terms of the Piola-Kirchhoff stress $\boldsymbol{\sigma}_P$, based on the current stress-free configuration.

While the elastic response of the material is given by an interpretation of relation (93) and provides directly the total stress, the inelastic constituent of the material has to be specified by a rate-type constitutive assumption. As a consequence, an integration in time of the appropriate inelastic evolution law is required which furnishes the inelastic strain entering in (93). In this connection, it should also be mentioned that the combination of a total- elastic and a rate-inelastic material constituent, in principle, leads to non-symmetric relations between stress rates and strain rates.

In certain situations the solid material can be modelled adequately as a purely inelastic viscous medium [26, 27]. The constitutive relation for an isotropic viscous material can be stated as

$$\boldsymbol{\sigma}_D = 2\mu\boldsymbol{\delta}_D \tag{96}$$

between the deviatoric part of Cauchy stress and the rate of deformation, and as

$$\sigma_H = 3\kappa\delta_V \tag{97}$$

between the hydrostatic (mean) stress and the volumetric rate fo deformation $3\delta_V$. The scalars μ and κ denote the deviatoric and the volumetric viscosity coefficient respectively. As a matter of fact, the constitutive relations in(96) and (97) contribute directly to the stress state $\boldsymbol{\sigma}$ and can be substituted immediately into the integral on the right-hand side of (1).

Usually, inelastic flow occurs under isochoric conditions for which $3\delta_V$ vanishes and relation (11) is not applicable. Nevertheless, a penalty approach to the isochoric condition makes use of (11) in the form

$$3\delta_V = \frac{1}{\bar{\kappa}}\sigma_H \rightarrow 0 \quad ; \quad \bar{\kappa} \rightarrow \infty \tag{98}$$

This facilitates the treatment of the incompressibility constraint by standard procedures. The penalty factor $\bar{\kappa}$ is chosen according to numerical aspects.

The deviatoric behaviour, still given by (96), requires a specification of the viscosity coefficient μ which is the decisive material characteristic. It may be a nonlinear function of the rate of deformation and of the accumulated strain if strain hardening effects are to be considered in addition to the viscous behaviour. From (96), a relation between the von Mises equivalent stress $\bar{\sigma}$ and the equivalent rate of deformation $\bar{\delta}$ can be deduced and reads

$$\bar{\sigma} = 3\mu\delta = f(\bar{\delta}, \bar{\gamma}) \tag{99}$$

The functional dependence indicated in (99) is usually derived from uniaxial tests, where

$$\bar{\gamma} = \int \bar{\delta} dt \tag{100}$$

serves as measure for the accumulated strain. From (99) the viscosity coefficient μ reads

$$\mu = \frac{f(\bar{\delta}, \bar{\gamma})}{3\bar{\delta}} = \mu(\bar{\delta}, \bar{\gamma}) \tag{101}$$

and may be specified in terms of the uniaxial material data.

Clearly, the case $\bar{\sigma} = f(\bar{\gamma})$ describes a solid which is rigid-plastic and then μ in (101) has not the physical significance of a viscosity coefficient. It is merely a fictitious one allowing the simulation of the rigid-plastic material by a viscous material, cf. [28, 29].

The integration of the rate constitutive law (92) with respect to time yields an expression for the stress in (1) in terms of strains. Similarly, the total relation (93) may be interpreted – after integration in time of the rate inelastic law – as a relation between stress and strain. The result of either procedure, which will in general involve an approximate integration with respect to time, may be symbolised by the dependence $\boldsymbol{\sigma}(\boldsymbol{\gamma})$. Furthermore, strains are determined from the deformed geometry of the solid as $\boldsymbol{\gamma}(\boldsymbol{X})$. Thus, ultimately, the deformed geometry is the only unknown appearing in the equilibrium condition as indicated in (11).

The solution for elastoplastic solids proceeds as outlined in section 2.2. The specification of the stress resultants and of the gradient matrices follows the lines of section 3.2 and section 3.3 respectively, but considers properly the material laws appertaining to large strains. Also, the geometric stiffness must *ab initio* account for changes in the reference geometry of the element due to the deformation.

In a viscous solid , the Cauchy stress is immediately expressed as a function of the rate of deformation by (96) and (97). As the rate of deformation derives from the actual velocity gradient, the stress resultants must be considered a function of the geometry and of the velocity as in the equilibrium condition (10). The computational implications of the viscous approach are discussed in the subsequent section.

4.2 The Viscous Approach

Isochoric Finite Elements

In the following, we are interested in the finite element formulation of the viscous isochoric deformation problem as based on the penalty approach. For this purpose we first refer, as usually, to the virtual work expression (1). Then, following the procedure in section 2.1, the finite element form (4) is derived on the basis of the interpolation (2) for the velocities $\boldsymbol{v}$ within the element and (3) for the rate of deformation $\boldsymbol{\delta}$.

In the penalty approach to the viscous isochoric solid, the stress $\boldsymbol{\sigma}$ is related to the rate of deformation $\boldsymbol{\delta}$ by (96) and (97), (98). Accordingly, the classical finite element formulation as based on the approximate velocity field yields the stresses

$$\boldsymbol{\sigma} = 2\mu\boldsymbol{\delta}_D + 3\bar{\kappa}\delta_V\boldsymbol{e} = [2\mu\boldsymbol{\alpha}_D + \bar{\kappa}\boldsymbol{e}\boldsymbol{\alpha}_V]\,\boldsymbol{V}_e \tag{102}$$

where the vector $\boldsymbol{e} = \{111000\}$ distributes $3\delta_V$ and $\boldsymbol{\alpha}_D, \boldsymbol{\alpha}_V$ denote the operator $\boldsymbol{\alpha}$ for the deviatoric, the volumetric rate of deformation respectively. Substitution of (102) for the stress in (4) furnishes the element stress resultants in terms of the element velocities in the form

$$\boldsymbol{P} = [\boldsymbol{d} + \bar{\boldsymbol{\kappa}}_V]\,\boldsymbol{V}_e \tag{103}$$

The matrix

$$\boldsymbol{d} = \int_V 2\mu\boldsymbol{\alpha}_D^t\boldsymbol{\alpha}_D \mathrm{d}V \tag{104}$$

is known as the deviatoric viscosity matrix of the element. The volumetric element matrix

$$\bar{\boldsymbol{\kappa}}_V = \int_V \bar{\kappa}\boldsymbol{\alpha}_V^t\boldsymbol{\alpha}_V \mathrm{d}V \tag{105}$$

results from the penalty formulation of the isochoric condition. In order to prevent overconstraint, the matrix $\bar{\boldsymbol{\kappa}}_V$ is evaluated by a reduced integration rule.

In a mixed finite element formulation, [30], an independent approximation is employed for the hydrostatic stress σ_H in the elements the hydrostatic pressure p respectively. We write,

$$-\sigma_H = p = [\pi_l]\{p_l\} = \boldsymbol{\pi p} \quad ; \quad l = 1, \ldots, r \tag{106}$$

The vector $\boldsymbol{p}$ comprises pivotal values of the pressure and $\boldsymbol{\pi}$ interpolation functions of a lower order than those for the velocities. The stress in the element is now written as

$$\boldsymbol{\sigma} = 2\mu\boldsymbol{\delta}_D - p\boldsymbol{e} = 2\mu\boldsymbol{\alpha}_D\boldsymbol{V}_e - \boldsymbol{e\pi p} \tag{107}$$

For a determination of $\boldsymbol{p}$ the penalty relation (98) is considered in the weak form

$$\int_V 3\bar{p}\delta_V \mathrm{d}V = -\int_V \frac{1}{\bar{\kappa}}\bar{p}\,p\,\mathrm{d}V \tag{108}$$

where $\bar{p}$ denotes a virtual pressure field. By introducing in (108) the interpolations (106) and (2) we deduce the finite element counterpart of (98)

$$\boldsymbol{h}^{\mathrm{t}}\boldsymbol{V}_e = -\bar{\boldsymbol{\kappa}}_H^{-1}\boldsymbol{p} \tag{109}$$

Here,

$$\boldsymbol{h} = \int_V \boldsymbol{\alpha}_V^{\mathrm{t}}\boldsymbol{\pi}\mathrm{d}V \tag{110}$$

is the hydrostatic matrix of the element, and

$$\bar{\boldsymbol{\kappa}}_H^{-1} = \int_V \frac{1}{\bar{\kappa}}\boldsymbol{\pi}^{\mathrm{t}}\boldsymbol{\pi}\mathrm{d}V \tag{111}$$

the penalty hydrostatic matrix which is different from the inverse of $\bar{\boldsymbol{\kappa}}_V$ in (105). We note *en passant* that for a vanishing right-hand side, (109) represents the finite element expression of the strict isochoric condition.

Solution of (109) for $\boldsymbol{p}$ and substitution in (107) yields the stress in terms of the element nodal velocities. Use of the resulting expression in (4) leads to the form

$$\boldsymbol{P} = \left[\boldsymbol{d} + \boldsymbol{h}\bar{\boldsymbol{\kappa}}_H\boldsymbol{h}^{\mathrm{t}}\right]\boldsymbol{V}_e \tag{112}$$

for the element stress resultants.

In either case, the velocity formulation or the mixed one, the stress resultants at the nodal points of the element may be written as

$$\boldsymbol{P} = \bar{\boldsymbol{d}}\boldsymbol{V}_e \tag{113}$$

The element viscosity matrix $\bar{\boldsymbol{d}}$ is defined by (103) for the velocity formulation and by (112) for the mixed formulation.

Computational Procedure

The standard assembly operation (7) leads from the element relation (113) to the form

$$\boldsymbol{S}(\boldsymbol{V},\boldsymbol{X}) = \bar{\boldsymbol{D}}(\boldsymbol{V},\boldsymbol{X})\boldsymbol{V} = \boldsymbol{R}(t,\boldsymbol{X}) \tag{114}$$

for the equilibrium condition (10) which governs quasistatic deformation of the viscous solid. The indicated dependence of the viscosity matrix $\bar{\boldsymbol{D}}$ of the system on the velocity reflects a nonlinear constitution of the viscous material. As a matter of fact, both the velocity $\boldsymbol{V}$ and the deformed geometry $\boldsymbol{X}$ appear as unknown quantities in (114). They are linked via an approximate integration of the velocity within a small interval in time $\tau = {}^bt - {}^at$. One obtains the relation

$${}^b\boldsymbol{X} = {}^a\boldsymbol{X} + (1-\zeta)\tau\,{}^a\boldsymbol{V} + \zeta\tau\,{}^b\boldsymbol{V} \tag{115}$$

with the parameter $0 \leq \zeta \leq 1$. When $\zeta = 0$, the approximate integration in (115) is explicit and requires the velocity at the beginning of the time interval. In this case, the governing equation (114) is stated at $t = {}^at$ for $\boldsymbol{V} = {}^a\boldsymbol{V}$ with $\boldsymbol{X} = {}^a\boldsymbol{X}$ known. When $\zeta > 0$, the integration is implicit and (114) is considered at $t = {}^bt$ with $\boldsymbol{X} = {}^b\boldsymbol{X} = \boldsymbol{X}\left({}^b\boldsymbol{V}\right)$ in accordance with (115) and requires a solution for $\boldsymbol{V} = {}^b\boldsymbol{V}$. In either case, (114) may be solved by the application of the iteration scheme

$$\boldsymbol{V}_{i+1} = \boldsymbol{V}_i + \boldsymbol{H}_i\left[\boldsymbol{R}_i - \boldsymbol{S}_i\right] \tag{116}$$

which is analogous to (12). The iteration matrix $\boldsymbol{H}_i$ is in most cases chosen in accordance with

$$\boldsymbol{H} = \bar{\boldsymbol{D}}^{-1} \tag{117}$$

which proves satisfactory. For an assessment of the convergence properties of the iteration procedure (116), we deduce for the difference between consecutive iterations the expression

$$\delta_{i+1}\boldsymbol{V} = [\boldsymbol{I} + \boldsymbol{HG}]\delta_i \boldsymbol{V} \tag{118}$$

where $\boldsymbol{I}$ denotes the identity matrix and $\boldsymbol{G}$ the gradient of the residuum $[\boldsymbol{R} - \boldsymbol{S}]$ with respect to the velocity $\boldsymbol{V}$. The requirement for convergence is based on the spectral norm of the magnification matrix in (118) and is given by (15).

An important contribution to the gradient $\boldsymbol{G}$ emanates from the stress resultants. Considering first the functional dependence $\boldsymbol{S}(\boldsymbol{\sigma}, \boldsymbol{X})$ of (9), we formally derive

$$\frac{\mathrm{d}\boldsymbol{S}}{\mathrm{d}\boldsymbol{V}} = \left[\frac{\partial \boldsymbol{S}}{\partial \boldsymbol{\sigma}}\right]_X \frac{\mathrm{d}\boldsymbol{\sigma}}{\mathrm{d}\boldsymbol{V}} + \left[\frac{\partial \boldsymbol{S}}{\partial \boldsymbol{X}}\right]_\sigma \frac{\mathrm{d}\boldsymbol{X}}{\mathrm{d}\boldsymbol{V}} \tag{119}$$

The first part of the second term in the above expression may be identified as the geometric stiffness matrix $\boldsymbol{K}_G$ of (20) appertaining to an inviscid solid, while from (115) we obtain the relation

$$\mathrm{d}\boldsymbol{X} = \zeta\tau\,\mathrm{d}\boldsymbol{V} \tag{120}$$

between the differentials.

As concerns the first term in (119), we observe that due to the appearance of the velocity gradient in (96), (98), the stress in a purely viscous solid may not only vary with the velocity, but also with the geometry. Consequently,

$$\frac{\mathrm{d}\boldsymbol{\sigma}}{\mathrm{d}\boldsymbol{V}} = \frac{\mathrm{d}\boldsymbol{\sigma}}{\mathrm{d}\boldsymbol{\delta}}\left[\frac{\partial \boldsymbol{\delta}}{\partial \boldsymbol{V}} + \frac{\partial \boldsymbol{\delta}}{\partial \boldsymbol{X}}\frac{\partial \boldsymbol{X}}{\partial \boldsymbol{V}}\right] = \left[\frac{\partial \boldsymbol{\sigma}}{\partial \boldsymbol{V}}\right]_X + \left[\frac{\partial \boldsymbol{\sigma}}{\partial \boldsymbol{X}}\right]_V \frac{\mathrm{d}\boldsymbol{X}}{\mathrm{d}\boldsymbol{V}} \tag{121}$$

The gradient of the stress resultants $\boldsymbol{S}$ with respect to the velocity $\boldsymbol{V}$ may ultimately be written as

$$\frac{\mathrm{d}\boldsymbol{S}}{\mathrm{d}\boldsymbol{V}} = \boldsymbol{D}_T + \boldsymbol{D}_G \tag{122}$$

The matrix

$$\boldsymbol{D}_T = \left[\frac{\partial \boldsymbol{S}}{\partial \boldsymbol{\sigma}}\frac{\partial \boldsymbol{\sigma}}{\partial \boldsymbol{V}}\right]_X = \left[\frac{\partial \boldsymbol{S}}{\partial \boldsymbol{\sigma}}\frac{\mathrm{d}\boldsymbol{\sigma}}{\mathrm{d}\boldsymbol{\delta}}\frac{\partial \boldsymbol{\delta}}{\partial \boldsymbol{V}}\right]_X \tag{123}$$

reflects the instantaneous viscous properties of the material [26] and is in the non-linear case different from the secant matrix $\bar{\boldsymbol{D}}$ in (114). The matrix

$$\boldsymbol{D}_G = \tau\zeta\left\{\left[\frac{\partial \boldsymbol{S}}{\partial \boldsymbol{\sigma}}\right]_X \left[\frac{\partial \boldsymbol{\sigma}}{\partial \boldsymbol{X}}\right]_V + \left[\frac{\partial \boldsymbol{S}}{\partial \boldsymbol{X}}\right]_\sigma\right\} \tag{124}$$

refers to the response of the stressed solid to variations in geometry.

4.3 Contact and Friction

Contact with the Die

In metal forming processes, deformation of the work-piece material is often constrained by a die, which is assumed to be rigid in the following. The tasks associated with the consideration of the unsteady contact between work-piece material and die in the numerical analysis concern mainly the specification of the surface limiting the motion of the material, the test for contact or penetration, and the accounting for the boundary conditions upon contact [32].

For general situations a discretised description of the geometry of the die surface is convenient. This is done by the use of individual surface elements in a mesh on the potential contact surface of the die. The surface limiting the motion of the work-piece material is then completely specified by the coordinates of the mesh nodal points, the topological description of the surface elements, and by the geometrical properties of the elements. In addition, it is useful to record the surface elements connected to each nodal point. The numbering order of the element nodal points defines the sign of the surface normal, which we assume to be positive by convention when pointing into the space prohibited for the material motion. A normal uniquely defined for each element simplifies the subsequent procedure and for this reason, plane elements are assumed in what follows.

The test for contact requires a comparison of the actual position of the work-piece material with that of the die. For this purpose let us consider a nodal point m on the surface of the discretised material, Fig. 4, and obtain first the nodal point k of the die surface closest to m by the condition of minimum distance,

$$\boldsymbol{x}^{t}_{km}\boldsymbol{x}_{km} = \min\left(\boldsymbol{x}^{t}_{jm}\boldsymbol{x}_{jm}\right) \quad ; \quad j = 1, \cdots, K \tag{125}$$

where $\boldsymbol{x}_{km}$ denotes the radius vector from k to m. At the beginning of the computation the search comprises each of the K nodal points of the discretised die surface. Later on, the procedure is first restricted to the vicinity of the previous neighbour k of m and is extended to other nodal points if necessary. Subsequently, the particular surface element e is specified which contains the projection

$$\boldsymbol{y}_{km} = \left[\boldsymbol{I} - \boldsymbol{n}_e\boldsymbol{n}^{t}_e\right]\boldsymbol{x}_{km} \tag{126}$$

of the vector $\boldsymbol{x}_{km}$ onto the plane with normal $\boldsymbol{n}_e$, Fig. 5a. The position of the point of the work-piece material relative to the surface of the die is ultimately indicated by the normal distance to this element,

$$d_n = \boldsymbol{n}^{t}_e\boldsymbol{x}_{km} \tag{127}$$

As long as $d_n < 0$, the nodal point has not reached the die and is left to move freely.

The numerical treatment of the phenomena occuring upon contact between the work-piece material and the die is based in the following on the viscous approach, as an example. When the material nodal point (velocity $\boldsymbol{v}$) contacts the die (velocity $\boldsymbol{w}$), the normal velocity $[\boldsymbol{v}_n - \boldsymbol{w}_n]$ directed into the die surface must be suppressed in the calculation, Fig. 5b. This implies a change in the kinematic boundary conditions for the work-piece material and affects the structure of the system matrix; it is therefore inconvenient. Alternatively, a penalty form of this contact condition reads

$$\boldsymbol{v}_n - \boldsymbol{w}_n = -\frac{1}{k_n}\boldsymbol{F}_n \rightarrow 0 \ ; \ k_n \rightarrow \infty \tag{128}$$

and helps to express the contact pressure $\boldsymbol{F}_n$ as a viscous force which contributes to the stress resultants in (114). In this manner the contact condition modifies the viscosity matrix of the system solely through a contribution by the penalty factor k_n, while the structure of the matrix remains unchanged.

When penetration of the die occurs the material point is brought back onto the die by a prescribed velocity normal to the die surface. For this purpose the approximate time integration of (115) is applied to the desired motion of the nodal point within the time increment

$$-d_n \boldsymbol{n}_e = (1-\zeta)\tau \ ^a[\boldsymbol{v}_n - \boldsymbol{w}_n] + \zeta\tau^b[\boldsymbol{v}_n - \boldsymbol{w}_n] \tag{129}$$

It yields the velocity $^a[\boldsymbol{v}_n - \boldsymbol{w}_n]$ for $\zeta = 0$, or the velocity $^b[\boldsymbol{v}_n - \boldsymbol{w}_n]$ for $\zeta > 0$. The calculated velocity is imposed on the respective node of the work-piece material by the penalty form (128) which is modified accordingly.

Friction Phenomena

Sliding motion of the work-piece material with local velocity $[\boldsymbol{v}_t - \boldsymbol{w}_t]$ along the surface of the die is opposed by a friction force $\boldsymbol{F}_t$, Fig. 5b. This can be expressed by the statement

$$\boldsymbol{F}_t = -\frac{|\boldsymbol{F}_t|}{|\boldsymbol{v}_t - \boldsymbol{w}_t|}[\boldsymbol{v}_t - \boldsymbol{w}_t] \tag{130}$$

which defines the direction of the friction force beyond sticking. The material sticks as long as $|\boldsymbol{F}_t| < F_H$, $F_H > 0$ being the sticking limit. For Coulomb-type friction,

$$|\boldsymbol{F}_t| \leq c|\boldsymbol{F}_n| \tag{131}$$

where the friction coefficient c may depend on various parameters of the process [29]. With (130), (131) the friction force can be determined using the quantities available in the course of the iteration process and contributes to the applied forces in (114), (116). Thereby a tendency to sticking is indicated by oscillations in the sliding velocity when approaching zero, and may lead to unreliable results. Stabilisation of the numerical results can be achieved by suppressing the sliding motion. This is an inconvenient step, because it affects the structure of the viscosity matrix of the system. An alternative is provided by the penalty formulation of the sticking condition [33], which is analogous to (128). It reads

$$\boldsymbol{v}_t - \boldsymbol{w}_t = -\frac{1}{k}\boldsymbol{F}_t \rightarrow 0 \ ; \ k \rightarrow \infty \tag{132}$$

However, the definition of an appropriate critical velocity ε for the appearance of sticking is crucial to both procedures, as is the definition of the penalty factor k in the latter, Fig. 6.

Expression (130) for the friction force and the penalty form of the sticking condition may be combined to a complete kinematic computational approach to friction [32]. The single form proposed reads

$$\boldsymbol{F}_t = -k_t\left[\boldsymbol{v}_t - \boldsymbol{w}_t\right] \tag{133}$$

where the factor k_t is defined as the ratio,

$$k_t = |\boldsymbol{F}_t| / |\boldsymbol{v}_t - \boldsymbol{w}_t| \leq k_{\max} \tag{134}$$

which represents the actual slope in Fig. 7. Limitation by $k_{\max}$ implies the penalty approach to the sticking condition. By means of equation (133), friction is accounted for throughout the numerical analysis by a viscous force in conformity with the contact pressure in (128), and contributes via k_t to the viscosity matrix of the system. The present approach to the friction force seems to be insensitive to the choice of the penalty parameter $k_{\max}$, and the velocity criterion to sticking.

The above approach to friction is studied on the thick-walled cylindrical specimen under compression shown in Fig. 8. A constant friction coefficient $c = 0.2$ is assumed at the interfaces with the die, and is by experience more severe for the calculation than one depending on the sliding motion [29]. The analysis is based on a rigid-plastic material law which complies with the uniaxial hardening characteristic in Fig. 8. At the ends of the cylinder, a prescribed axial velocity simulates the motion of the die such that a reduction in height by 16 mm is attained with 100 equal increments in time (explicit, $\zeta = 0$). For illustration, two stages of the deformation process are depicted in Fig. 9.

In a first investigation, friction is accounted for via the application of forces, sticking by the penalty approach. The penalty factor $k = 10^{-9}$ appears to be too high in the present case and remains inactive in the procedure. As a consequence, the numerical oscillations related to sticking occur at low sliding velocities and are reflected by the results presented in Fig. 10. The radial velocity V_r along the contact surface is a quantity sensitive to friction and is shown in the figure as a function of the reduction in height Δh of the cylinder. The numerical difficulties increase as the region of low sliding velocities increases. The results become meaningless and are therefore reproduced for the first 73 of the 100 time increments employed.

A second study makes use of the kinematic approach to friction as by (133) which contributes to the viscosity matrix of the system via the factor k_t, (134), with a limitation by $k_{max} = 10^{-9} = k$. In this case the previous oscillations do not appear and the results shown in Fig. 11 are smooth. Thereby, also the insensitivity with respect to k_{max} is confirmed.

Finally, the alternative elastoplastic methodology should be considered for completeness. In this context kinematics are described in terms of displacements which replace the velocity variables of the viscous approach. Then, the penalty factor k_n for the normal contact force in (128), as well as the factor k_t for the friction force in (133), refer to relative displacements between the work-piece material and the die and contribute, via the internal stress resultants in (10), to the gradient matrix (14) of the elastoplastic system.

4.4 Adaptive Mesh Refinement

The demand for a variation of the finite element discretisation arises from the necessity of obtaining an optimal numerical solution to a given problem. When the optimal computational mesh cannot be designed in advance, a trial-and-error approach may be employed [34]. Thereby a first solution is obtained with a trial discretisation; analysis of the quality of the solution indicates necessary improvements and the mesh is redesigned in compliance with certain posed requirements. Modifications to the mesh may be performed either by displacing the nodal points or by subdivision of elements – or by a change of the approximation order in the element [35].

In the simulation of large deformation processes, an initially reasonable discretisation of the material often deteriorates due to severe deformations in the course of the calculation. Accordingly, the numerical solution loses its quality. Regeneration of the computational mesh is based here on a displacement of the nodal points relative to the material within the common boundary. Thereby the size of the problem and the structure of the data remain unaltered. This is of some importance in connection with the matrix method of solution used in the computation algorithm. The incremental solution of the large deformation problem advances the result of previous calculations, which is required at the beginning of each incremental step.

Therefore, when the mesh is regenerated the solution for the existing mesh must be referred to the new mesh, cf. [32].

For this purpose, the so-called mixed Eulerian-Lagrangean technique considers a motion of the computational mesh relative to the material. A nodal point r coinciding at the beginning of the time increment with the material point s at location $\boldsymbol{x}$, moves with velocity $\boldsymbol{w}$, the velocity of the material point is $\boldsymbol{v}$, Fig. 12. The change with time of a substantial field quantity $q(\boldsymbol{x}, t)$ in the computational mesh reads then

$$\dot{q}_r = \dot{q}_s + \frac{\partial q}{\partial \boldsymbol{x}}[\boldsymbol{w} - \boldsymbol{v}] \tag{135}$$

Here, $\dot{q}_r$ denotes the time rate of q at the mesh nodal point $r, \dot{q}_s$ denotes the time rate at the material point s as governed by the deformation of the material. The convective transfer of field quantities as given by (135) is based on the respective gradient which represents a critical term in discretised solutions. Furthermore, the incremental technique deriving from (135) presumes small displacements of the mesh relative to the material and is usually employed in conjunction with a continuous variation of the mesh during the course of the calculation.

In an alternative procedure the solution is referred to the new mesh by means of an interpolation. To this end assume the field quantities at the nodal points of the existing mesh to be known. The new position of the nodal points after regeneration is given by the respective coordinates. In addition, the transfer of the solution requires a specification of the varied positions within the existing mesh. In analogy to the technique outlined in section 4.3 for the contact between material and die, we obtain for each nodal point r in the new mesh first the nodal point s in the existing mesh closest to r. Subsequently, the particular element e connected with s is found which contains the new nodal point r, Fig. 13. The field quantities at r are then obtained in accordance with the finite element approximation

$$q_r = \boldsymbol{\omega}_e(\boldsymbol{x}_r)\boldsymbol{q}_e \tag{136}$$

Here, the vector array $\boldsymbol{q}_e$ comprises the values at the nodal points of the element e, and $\boldsymbol{\omega}_e$ contains the interpolation functions. The value q_r at nodal point r in the new mesh is obtained by evaluating the interpolation for the position $\boldsymbol{x}_r$. Conversely to (135), the present technique allows for an occasional regeneration of the mesh. Furthermore, the number of nodal points in the two meshes may be different.

The interpolation algorithm is employed in the study of the axisymmetric tensile specimen in Fig. 14. A small geometric imperfection initiates necking in the middle region of the specimen. The material constitutive law is assumed rigid-plastic. The discretisation shown is coarse. In the calculation (explicit, $\zeta=0$) a relative elongation of 0.25×10^{-2} is imposed within each time increment. Results are obtained with two alternative meshes. One is embedded in the material and deforms with it, the other is regenerated after each fifth time increment. The regeneration of the mesh is adapted to the development of the deformation, so that the elements in the center of the specimen remain nearly quadratic while the element length increases gradually towards the end of the specimen. Throughout the calculation the size and the structure of the problem are kept constant. Figure 15 presents results of the two meshes for a relative elongation of 0.35. The mesh deforming with the material obviously does not favour the development of necking as the repeatedly regenerated mesh does. Also, the deformation of the material obtained with the adaptive mesh exhibits marked deviations from the deforming material mesh. Figure 16 demonstrates the concentration of the adaptive mesh in the necked region during the course of the calculation. Figures 17 and 18 show the development of necking and the variation of the tensile force, respectively, compared for the two meshes. In both cases the differences between the material and the adapted mesh are significant.

Alternatively, the movement of the nodal points may be controlled by an error criterion. An error estimate proposed in [36] for the viscous approach to forming problems is based on the energy norm of error,

$$\|e\|^2 = \int_V [\boldsymbol{\sigma}_D - \boldsymbol{\sigma}_D^*]^t \, \frac{1}{2\mu} \, [\boldsymbol{\sigma}_D - \boldsymbol{\sigma}_D^*] \, \mathrm{d}V \tag{137}$$

Here, the deviatoric stress $\boldsymbol{\sigma}_D$ results directly from the computation, and $\boldsymbol{\sigma}_D^*$ from a subsequent smoothing procedure; μ denotes the viscosity coefficient. The adaptive regeneration procedure displaces nodal points so that the error measure of (137) remains constant over an optimal mesh layout.

5. Applications

5.1 Crash Analysis

Nonlinear Dynamics

The equations governing under crash conditions the impact motion and deformation of the structure discretised by finite elements assume the matrix form

$$\boldsymbol{R}(t,\boldsymbol{X}) = \boldsymbol{S}(\boldsymbol{X},\boldsymbol{V}) + \boldsymbol{M}\dot{\boldsymbol{V}} \tag{138}$$

In (138), the inertia forces are obtained with the consistent mass matrix $\boldsymbol{M}$ of the structure. The stress resultants $\boldsymbol{S}$ may depend, according to the material constitution, on the deformed geometry $\boldsymbol{X}$ and on the velocity $\boldsymbol{V}$ in the case of a viscous material.

A fully algebraic equation is obtained by linking in (138) geometry, velocity and acceleration via an approximate integration within a time increment $\tau = {}^bt - {}^at$. To this end we introduce the following implicit approximation scheme, which is based on the acceleration ${}^b\dot{\boldsymbol{V}}$, at the end of the considered time increment. It furnishes the velocity

$$ {}^b\boldsymbol{V} = {}^a\boldsymbol{V} + \tau \left[a\,{}^a\dot{\boldsymbol{V}} + b\,{}^b\dot{\boldsymbol{V}} \right] \tag{139} $$

and the geometry

$$ {}^b\boldsymbol{X} = {}^a\boldsymbol{X} + \tau\,{}^a\boldsymbol{V} + \tau^2 \left[c\,{}^a\dot{\boldsymbol{V}} + d\,{}^b\dot{\boldsymbol{V}} \right] \tag{140} $$

at time instant ${}^bt$, the end of the increment. The parameters a, b, c, d in (139) and (140) control the numerical performance of the approximate integration and are chosen accordingly. They may be adapted to particular time stepping schemes already known from the pertinent literature [38, 39, 20], cf. also [40].

With the aid of (139) and (140) either the deformed geometry $\boldsymbol{X} = {}^b\boldsymbol{X}$ or the velocity $\boldsymbol{V} = {}^b\boldsymbol{V}$, or the acceleration $\dot{\boldsymbol{V}} = {}^b\dot{\boldsymbol{V}}$, may be considered to constitute the unknown variables in (138). The problem of dynamic motion and deformation can therefore be solved at time instant $t = {}^bt$ for one of the variables $\boldsymbol{X}, \boldsymbol{V}$ or $\dot{\boldsymbol{V}}$ at the end of the time increment, the other variables then deriving from the approximate scheme (139), (140).

As an example, consider in the following the acceleration $\dot{\boldsymbol{V}} = {}^b\dot{\boldsymbol{V}}$ as the independent variable. A solution of the nonlinear system (138) is given on the basis of the iterative procedure of (12). In the present dynamic case the relevant recurrence formula reads

$$ \dot{\boldsymbol{V}}_{i+1} = \dot{\boldsymbol{V}}_i + \boldsymbol{H}_i \left[\boldsymbol{R}_i - \boldsymbol{S}_i - \boldsymbol{M}\dot{\boldsymbol{V}}_i \right] \tag{141} $$

and furnishes the result of iteration $i+1$ using the data obtained in the ith iteration cycle.

One may observe that the residual $[\boldsymbol{R} - \boldsymbol{S} - \boldsymbol{M}\dot{\boldsymbol{V}}]$ in (141) differs from the quasistatic one in (12) or (116) merely by the inertia term $\boldsymbol{M}\dot{\boldsymbol{V}}$. Accordingly, when the gradient of the system is formed for an assessment of the iteration matrix $\boldsymbol{H}$ one obtains,

$$ \boldsymbol{G} = \frac{\mathrm{d}}{\mathrm{d}\dot{\boldsymbol{V}}} \left[\boldsymbol{R} - \boldsymbol{S} - \boldsymbol{M}\dot{\boldsymbol{V}} \right] = \frac{\mathrm{d}}{\mathrm{d}\dot{\boldsymbol{V}}} [\boldsymbol{R} - \boldsymbol{S}] - \boldsymbol{M} \tag{142} $$

Furthermore, with reference to the approximation by (139), (140),

$$\frac{\mathrm{d}}{\mathrm{d}\dot{\boldsymbol{V}}}[\boldsymbol{R}-\boldsymbol{S}] = \tau b \frac{\mathrm{d}}{\mathrm{d}\boldsymbol{V}}[\boldsymbol{R}-\boldsymbol{S}] = \tau^2 d \frac{\mathrm{d}}{\mathrm{d}\boldsymbol{X}}[\boldsymbol{R}-\boldsymbol{S}] \tag{143}$$

where the second and third expression are built with the gradient matrices related to the viscous quasistatic system in (116) and the inviscid elastoplastic system in (12) respectively, cf. (14). The complete gradient of the dynamic problem in (142) requires the addition of the consistent mass matrix $\boldsymbol{M}$ to the above terms.

A pure vector iteration procedure for the solution of the nonlinear dynamic equations of motion arises when the inverse of the diagonal lumped mass of the structure is used as an iteration matrix in (141). Of course, the residual vector is computed entirely on the basis of contributions from individual elements and does not require an assembly of the consistent mass matrix of the structure. Also, the time marching scheme is explicit in the case where the velocity given by (139) and the geometry given by (140) at the end of the time increment do not depend on the acceleration at that time instant.

Crash of Front Car Structure

The following investigation (cf. [37]) deals with the crash of a typical car structure. The empty front structure of the car is fitted to a heavy mass, representing the remainder of the structure and is struck against a rigid barrier, as indicated in Fig. 19. The total impact mass of the assembly is $M = 588$ kg, the impact velocity is $V = 49.3$ km/h. Two different steel sheet materials are employed in the car structure. A high-strength steel sheet is used in the longitudinal main beams – which can be recognised in Fig. 19 – while the remainder of the front structure consists of steel sheet satisfying current standards in the car manufacturing industry.

The behaviour of the two materials is obtained from uniaxial tensile tests, and is presented for the standard steel sheet in Fig. 20; the characteristic of the high-strength material is similar. In both cases, the engineering stress-strain diagrams indicate a dependence on the strain rate between the lower and upper limiting strain rates indicated in the figure. Thus, the materials do not exhibit remarkable rate effects below an engineering strain rate of about $0.05\mathrm{s}^{-1}$ and above a limit of about $10.0\mathrm{s}^{-1}$.

Figure 21 may give an impression of the discretisation of the car structure by finite elements. Assuming the crash problem to be symmetrical about the longitudinal vertical midplane, only half of the shown configuration is considered in the computation. The finite element mesh then consists of 1194 nodal points comprising 2293 simple triangular plate elements (3 nodes), together with 36 beam elements, both modified for purely inelastic viscous behaviour. As a result of the assumed symmetry and of the rigid connection of the structure with the end plate, the total number of unknowns amounts to 6731 in the crash problem.

The crash analysis starts with the impact velocity taken as the initial velocity of the entire discretised structure plus additional mass, except for the positions which collide with the rigid barrier first and therefore have zero velocity. The material constitutive model is assumed to be rigid-viscous as described by (96), (98). Strain hardening and rate effects are accounted for via the viscosity coefficient specified on the basis of (101) and the experimental test data.

Integration in time of the equations of motion follows an adaptation of Newmark's 'unconditional stable' algorithm [38] for the integration scheme of (139), (140). Then, $a = b = 1/2$ in (139) and $c = d = 1/4$ in (140). The variable in the equations of motion was identified with the acceleration $\dot{\boldsymbol{V}}$, and the solution by (141) was based on an iteration matrix $\boldsymbol{H}$ representing the inverse of a close approximation to the gradient matrix $\boldsymbol{G}$ of the system. During the course of the computation, the increments of time typically varied from 0.1 ms to 0.2 ms.

Results of the computer simulation of the crash are presented in Figs. 22 to 24. Deformation leads first to a flattening of the bumper; subsequently the main longitudinal beams are loaded by contact with the barrier. Initially, the main beams are deformed locally along the longitudinal axis and folds are formed as shown in Fig. 22. In this figure, the front structure is shown from above in the undeformed configuration and in a deformed state after a forward motion of 191.7 mm. Further deformation leads to an overall buckling of the main beam. This buckling initiates a downward movement of the beam, as shown in the perspective view of Fig. 23, which corresponds to a front end displacement of 246.5 mm. Figure 24 shows the final stage of deformation of the crashed structure from the side, as obtained numerically for an overall forward notion of 301.2 mm of the front end.

5.2 Simulation of Hot-Forging Processes

Thermomechanical Coupling

In the non-isothermal case, the temperature in the deforming solid appears as an additional variable of the process and is governed by the energy balance. The finite element formulation of the thermal problem leads to the matrix equation.

$$\dot{\boldsymbol{Q}} - \boldsymbol{C}\dot{\boldsymbol{T}} - \boldsymbol{L}\boldsymbol{T} = \boldsymbol{o} \tag{144}$$

where $\dot{\boldsymbol{Q}}$ is the heat flux vector, $\boldsymbol{C}$ denotes the heat capacity matrix and $\boldsymbol{L}$ the heat conductivity matrix of the discretised solid material. Thermal effects can arise in the deforming solid by externally applied thermal action and irreversible, as well as reversible, deformations. As a consequence, $\boldsymbol{L}$ is extended to account for reversible mechanical effects, dissipative mechanical contributions are considered in $\dot{\boldsymbol{Q}}$, [41]. Furthermore, equation (144) requires knowledge of the actual geometry of the solid, while the temperature appears in the stress-strain relations for the material [29, 31]. Therefore, thermal and mechanical phenomena are coupled.

When the mechanical coupling terms in (144) are assumed frozen the unknown quantities are $\dot{\boldsymbol{T}}$ and $\boldsymbol{T}$. They are linked via an approximate integration of the temperature rate within the small time interval $\tau = {}^bt - {}^at$, which yields the relation

$$^b\boldsymbol{T} = {}^a\boldsymbol{T} + (1-\zeta)\tau\, {}^a\dot{\boldsymbol{T}} + \tau\zeta\, {}^b\dot{\boldsymbol{T}} \tag{145}$$

with the parameter $0 \leq \zeta \leq 1$. When $\zeta = 0$, the approximate integration in (145) is explicit and requires the temperature rate at the beginning of the time interval. In this case, equation (144) is stated at $t = {}^at$ for $\dot{\boldsymbol{T}} = {}^a\dot{\boldsymbol{T}}$ with $\boldsymbol{T} = {}^a\boldsymbol{T}$ known. When $\zeta > 0$, the integration is implicit and (144) is considered at $t = {}^bt$ with $\boldsymbol{T} = {}^b\boldsymbol{T} = \boldsymbol{T}({}^b\dot{\boldsymbol{T}})$ in accordance with (145) and requires a solution for $\dot{\boldsymbol{T}} = {}^b\dot{\boldsymbol{T}}$.

The coefficient matrices $\boldsymbol{C}$ and $\boldsymbol{L}$ in (144) as well as the vector $\dot{\boldsymbol{Q}}$ may depend also on the temperature $\boldsymbol{T}$ which leads, in the implicit case $\zeta > 0$, to an nonlinear expression for $\dot{\boldsymbol{T}}$. In this case, the iterative scheme of (12) as applied to the solution of (11) may analogously be used to solve the residual form (144) governing the transient temperature distribution in the solid. The same scheme is also applicable to the explicit case $\zeta = 0$.

In the elastoplastic case, the coupled thermomechanical problem is governed by the equilibrium condition (11) and the thermal equation (144). A solution technique for the coupled system of equations is based on the sequential treatment of the individual problems [41, 42]. Thereby, the equilibrium condition (11) is fulfilled first and yields the deformed geometry $\boldsymbol{X}$, while the estimated distribution of temperature is held fixed, $\boldsymbol{T}=$ const. . The computed mechanical variables are subsequently used in the thermal equation (144) which is solved for $\boldsymbol{T}$ while $\boldsymbol{X}=$ const., and the iteration cycle is repeated.

In the viscous approach, the equilibrium condition is used in the form of (114), while temperature is still governed by a suitable specification of (144). A solution of the arising coupled thermomechanical system is achieved along the lines discussed above for elastoplastic solids.

Forging of a Compressor Blade

In the following we consider the forging process of a compressor blade [27]. The initial geometry of the work-piece material and the geometry of the final product are specified in Fig. 25. One recognises the complex three-dimensional nature of the forging process involving unsteady contact of material and die. In addition, the significant coupling between thermal and mechanical phenomena must be accounted for during the course of the hot-forging process.

Figure 26 gives a schematic description of the forging process. The hot material, actually Ti-6Al-4V initially at ${}^0T = 1203$ K, is formed through a relatively cold die with $T_\infty = 423$ K. In order to prevent extensive heat loss the die velocity must be high, here taken to be $V = 0.1$ m/s. Heat transfer occurs to the surrounding air and to the die as respective surfaces are in contact. The relevant heat transfer coefficients are specified in the figure. Also of importance are heat generation due to dissipation within the work-piece material and heat conduction. The contact between work-piece material and die is assumed to be frictionless.

The thermomechanical properties of the Ti-6Al-4V alloy in the range of interest are indicated in Fig. 27. Under the assumed forging conditions this material may be assumed to behave as a rigid-viscous solid. The dependence of the uniaxial stress σ on the rate of deformation $\dot{\delta}$ is shown in the diagram on the left-hand side of Fig. 27 for different temperatures covering the range of interest. The diagram on the right-hand side of this figure specifies the dependence of the thermal conductivity λ of the material and of its specific heat capacity c on the temperature T.

The finite element discretisation of the initial work-piece material is shown in Fig. 28. The mesh comprises 1356 nodal points, 1020 hexahedral eight-node elements. In conjunction with the penalty approach the volumetric response of the viscous element is obtained by a reduced integration rule. The problem requires the treatment of 3676 unknown velocities and 1356 unknown temperatures necessary for the description of the unsteady thermomechanically coupled process.

The kinematic boundary conditions for the work-piece material are updated in the course of the computation according to the motion of the die towards the desired ultimate shape of the blade. In the present case, the geometry of the die surface can be described analytically. Concerning the foot of the component, out-of-plane motion is suppressed on the front and back vertical surfaces while the sides are free to move. The motion of the upper and lower horizontal planes has to follow the die.

The numerical simulation of the forging process is based on a thermomechanically coupled computation. The deformation process is governed by the mechanical equation (114) appertaining to the rigid viscous approach. The transient distribution of the temperature in the solid is governed by the thermal equation (144) which is coupled to the mechanical equation via the deforming geometry of the solid and the mechanical dissipative phenomena. The forging process is traced numerically using 100 equal increments of time $\tau = 2.5$ ms. Results of the computation are shown in Figs. 29 and 30. Figure 29 illustrates stages of deformation during the course of the forging process at 80%, 90% and 100% of the ultimate die motion. The associated temperature distribution is given in Fig. 30 for the surface of the work-piece material. A temperature decrease of approximately 400 K is caused by the contact with the cold die.

5.3 Superplasticity

Constitutive Description

The term superplasticity is used for deformation processes producing high, essentially neck-free permanent elongations in metallic materials subject to tension. Superplastic deformation requires a microstructure in the material exhibiting a stable, ultra-fine grain size at a temperature of deformation $T \geq 0.4T_M$, T_M denoting the absolute melting temperature of the material [43].

Fundamental to superplasticity is the dependence of the uniaxial flow stress σ on the rate of deformation δ, which prevents the formation of localised necks during the course of the flow process. In addition, the flow stress depends on the grain size d of the material, Fig. 31. Accordingly, one may write for isothermal conditions the functional dependence

$$\sigma = f(\delta, d) \tag{146}$$

for the flow stress. Superplasticity requires a certain amount of rate sensitivity and is therefore possible within a certain range of the rate of deformation. The uniaxial constitutive description is completed by a law for the evolution of the grain size in time,

$$\dot{d} = g(t, \delta) \tag{147}$$

which reflects the grain growth kinetics of the material, cf. Fig. 32.

The constitutive description under multiaxial conditions may be based on the assumption of a nonlinear viscous isochoric material [26] and is given by (96) and (98). For a specification of the viscosity coefficient μ in (96), we compare the relation (99) between the von Mises equivalent stress $\bar{\sigma}$ and rate of deformation $\bar{\delta}$ with the uniaxial relation (146) which is interpreted for the equivalent quantities. This yields

$$\bar{\sigma} = 3\mu\bar{\delta} = f(\bar{\delta}, d) \tag{148}$$

from which the viscosity coefficient derives as

$$\mu = \frac{f(\bar{\delta}, d)}{3\bar{\delta}} = \mu(\bar{\delta}, d) \tag{149}$$

in terms of uniaxial material data.

The forming process considered subsequently requires a specification of the constitutive law for the Ti-6Al-4V alloy at a temperature $T = 1200$ K. For this purpose we use the data reported in [44]. A suitable approximation for the uniaxial test data for the flow stress in Fig. 31 is provided by a power series expression yielding

$$\log \bar{\sigma} = a_0 + a_1 \log \bar{\delta} + a_2(\log \bar{\delta})^2 + a_3(\log \bar{\delta})^3 + a_4(\log \bar{\delta})^4 \tag{150}$$

The evolution law for the grain size is deduced from the data in Fig. 32 as

$$\frac{\dot{d}}{{}^0d} = \frac{N}{{}^0t}\left(\frac{t}{{}^0t}\right)^{N-1} \quad ; \quad N = 1.8(\bar{\delta} + 0.00005)^{0.237} \tag{151}$$

where ${}^0d$ denotes the initial grain size at time ${}^0t$.

Superplastic Forming

Superplastic forming allows the production of structural components with complex geometry requiring high permanent deformations in a single fabrication cycle. Due to the associated savings in weight and material this manufacturing process appears particularly attractive to the air- and spacecraft industry [26].

In the design phase of a superplastic forming process several tasks must be performed. They concern, on the one hand, the conception of the die and the initial state of the material to be formed; on the other hand, when carrying out the process, the variation of forming pressure must, as a rule, minimise time while maintaining the property of superplasticity. The final product must fulfill prescribed requirements with respect to its shape and strength. For certain reasons, trial forming processes are often not feasible and have to be replaced by computational experiments. The computation accompanying a careful design in superplastic forming becomes thus decisive for the realisation of the fabrication phase [45].

Superplastic forming processes are usually performed by the application of gas pressure. The numerical analysis presumes the knowledge of the time history of the forming pressure. This information is, however, not known in the design phase of the process. Moreover, the variation of the forming pressure with time is one of the numerical results requested. A minimum process time is to be achieved in compliance with the superplastic conditions which ensure that the material will not fail by the formation of localised necks. To this end the maximum rate of deformation within the forming component must not exceed a certain value associated with the maximum rate sensitivity of the flow stress characteristic for the material. In conformity with the above requirements the numerical simulation must be controlled by the optimal value of the maximum rate of deformation; the intensity of the forming pressure is to be adjusted accordingly. A general procedure for the adaptation of the forming pressure during the course of the computation may be based on a trial-and-error approach.

Consider, as an example, a statically determined system. At a certain stage of the forming process, application of the pressure loading with a trial intensity p, results in the distribution $\boldsymbol{\sigma}(\boldsymbol{x})$ of the stresses in the material. Let p_o denote the unknown optimal value of the pressure at this stage and $\boldsymbol{\sigma}_o(\boldsymbol{x})$ the associated distribution of the stresses. Static determinacy implies the validity of the relation

$$\boldsymbol{\sigma}_o(\boldsymbol{x}) = \frac{p_o}{p}\boldsymbol{\sigma}(\boldsymbol{x}) \tag{152}$$

which transfers also to the von Mises equivalent stresses

$$\frac{\bar{\sigma}_o(\boldsymbol{x})}{\bar{\sigma}(\boldsymbol{x})} = \frac{p_o}{p} \tag{153}$$

everywhere within the material. The maximum equivalent rate of deformation $\bar{\delta}_{\mathbf{max}}$, which is obtained at location $\boldsymbol{x}_{\mathbf{max}}$, must attain the prescribed value δ_o. Thus we may write

$$\frac{\bar{\sigma}_o(\boldsymbol{x}_{\mathbf{max}})}{\bar{\sigma}(\boldsymbol{x}_{\mathbf{max}})} = \frac{f(\delta_o)}{f(\bar{\delta}_{\mathbf{max}})} \tag{154}$$

Here, $f(\delta)$ refers to (146) taken with the actual grain size d. Comparison of (153) with (154) yields for the optimal pressure

$$\frac{p_o}{p} = \frac{f(\delta_o)}{f(\bar{\delta}_{\mathbf{max}})} = \left(\frac{\delta_o}{\bar{\delta}_{\mathbf{max}}}\right)^m \tag{155}$$

The last expression in (155) presumes the special constitutive form

$$\sigma = k\delta^m = f(\delta) \tag{156}$$

which is frequently quoted in the literature. It represents in fact only a linear approximation to the actual logarithmic curves reproduced in Fig. 31 and is merely locally appropriated.

If the constitutive law (156) is nevertheless assumed to be uniquely valid within the forming component, its use in (153) yields for the equivalent rate of deformation

$$\frac{\bar{\delta}_o(\boldsymbol{x})}{\bar{\delta}(\boldsymbol{x})} = \left[\frac{\bar{\sigma}_o(\boldsymbol{x})}{\bar{\sigma}(\boldsymbol{x})}\right]^{1/m} = \left(\frac{p_o}{p}\right)^{1/m} \tag{157}$$

Accordingly, the two distinct distributions of the complete rate of the isochoric deformation are related by

$$\boldsymbol{\delta}_o(\boldsymbol{x}) = \left(\frac{p_o}{p}\right)^{1/m} \boldsymbol{\delta}(\boldsymbol{x}) \tag{158}$$

which may be compared with relation (152) for the associated stress distributions.

Three-Dimensional Structural Component

As a selected application of the computational methodology developed, we consider in the following the superplastic forming of a three-dimensional structural component from a plane sheet. The plane sheet (448 mm × 428 mm) , shown discretised in Fig. 33, is formed onto a three-dimensional die by gas pressure. The discretised description of the die surface is shown in Fig. 34. Due to the symmetry only one half of the problem is considered as indicated. This complex investigation is related to the design of an industrial process of superplastic forming [45].

The sheet from Ti-6Al-4V material is formed at the temperature $T = 1200$ K. The constitutive behaviour of the material is characterised by relations (150) and (151), where the initial grain size is assumed to be ${}^0d = 8\mu$m at time ${}^0t = 600$ s. The initial thickness of the sheet is 3 mm.

In the computer simulation, the sheet is modelled by simple triangular plate elements. For this purpose an originally elastic element has been adapted to superplasticity [32]. Under the condition of plane stress as it appears in thin sheet forming processes, for instance, the isochoric constraint does not affect the stress-rate of deformation relations. It is merely used to update the sheet thickness in conformity with the in-plane deformation.

The plane stress vector is defined as

$$\boldsymbol{\sigma} = \{\sigma_{11} \;\; \sigma_{22} \;\; \sqrt{2}\sigma_{12}\} \tag{159}$$

and the corresponding rate of deformation vector is

$$\boldsymbol{\delta} = \{\delta_{11} \;\; \delta_{22} \;\; \sqrt{2}\delta_{12}\} \tag{160}$$

Application of the three-dimensional constitutive relation for the viscous material (96) yields for the plane stress under consideration

$$\boldsymbol{\sigma} = \boldsymbol{\sigma}_D + \boldsymbol{\sigma}_H = 2\mu\boldsymbol{\delta} + \sigma_H \boldsymbol{e} \tag{161}$$

Here the reduced vector $\boldsymbol{e} = \{110\}$ appertains to the two-dimensional case. In addition to (161) the isochoric constraint expresses the rate of deformation across the thickness t in terms of the in-plane components as

$$\frac{\dot{t}}{t} = -\boldsymbol{e}^{\mathrm{t}}\boldsymbol{\delta} \tag{162}$$

The plane stress state implies the compensation of the deviatoric component normal to the plane by the hydrostatic stress σ_H. The deviatoric stress is related to the rate of deformation by the viscosity of the material. Observing also (162) we obtain for the hydrostatic stress

$$\sigma_H = 2\mu\boldsymbol{e}^{\mathrm{t}}\boldsymbol{\delta} \tag{163}$$

Substitution of (163) in (161) yields the complete constitutive relation for an isochoric viscous material in plane stress

$$\boldsymbol{\sigma} = 2\mu\left[\boldsymbol{I} + \boldsymbol{e}\boldsymbol{e}^{\mathrm{t}}\right]\boldsymbol{\delta} \tag{164}$$

The above expression can also be derived in direct analogy to an elastic incompressible material in plane stress with shear modulus μ and the rate of deformation in place of the elastic small strain. By this analogy, standard elastic finite elements may be easily utilised for the present purposes. The elastic analogy is, of course, not restricted to plane stress. However, three-dimensional situations require a particular imposition of the incompressibility constraint. Therefore, only special incompressible elastic elements might be considered for isochoric viscous behaviour in this case.

The process of forming is performed at constant temperature by gas pressure. In the calculation the intensity of the pressure is adjusted to a prescribed value for the maximum rate of deformation $\delta_o = 3 \times 10^{-4}\mathrm{s}^{-1}$ by a trial-and-error approach. The simple relation

$$\frac{p_o}{p} = \frac{\delta_o}{\bar{\delta}_{\max}} \tag{165}$$

is used instead of (155) in this connection and appears satisfactory in the problem considered. A limitation of the quotient in (165) helps to avoid strong numerical oscillations in the pressure. Time marching follows an explicit integration ($\zeta = 0$) with increments increasing from initially $\tau = 0.05$ s to ultimately $\tau = 30$ s in the early phase of the process. Upon contact the material is assumed to slide along the die without friction.

Figure 35 shows the deformed mesh at the end of the calculation, $t = 6410$ s. The development of the deformation of the finite element mesh is depicted in Fig. 36. The requested part may be considered formed at $t = 6410$ s. At this stage, the thickness of the sheet is reduced from initially 3 mm beyond the prescribed tolerance to 1.28 mm locally, and the grain size varies between 13.2μm and 15.6μm (initially 8μm). It can be seen from the numerical results that the maximum grain size appears at the same location as the minimum thickness. Figure 37 shows the variation of the forming pressure with time under observance of the given maximum rate of deformation $\delta_o = 3 \times 10^{-4}\mathrm{s}^{-1}$, cf. Fig. 38. Peaks usually occur when the material is brought back onto the die surface upon the occurence of penetration.

It should be recalled that the present study neglects friction between the sheet and the die. An inclusion of friction phenomena might lead to different conclusions.

5.4 Damage of Elastoplastic Materials

Constitutive Concept

Failure phenomena in ductile metals are attributed to the formation of microscopic voids, mainly from inclusions, their growth and coalescence during the course of plastic deformation in the material [46]. Accordingly, a concept for the constitutive description of a ductile metal, including damage and failure may consist in the main of a yield condition modified to account for the porosity in the material [47], in conjunction with a criterion for the appearance and an evolution law for damage [48].

Consider an isotropic solid which undergoes elastoplastic deformation and damage due to the nucleation and growth of microscopic voids within the basic matrix material. It is attempted in the following to describe the behaviour of such a solid material by a suitable modification of standard elastoplastic stress–strain relations [49]. To this end the local void volume fraction

$$f = V_v/V \quad ; \quad 0 \le f \le 1 \tag{166}$$

is introduced as an additional variable and relates the instantaneous void volume V_v to the current volume V of the material element. Accordingly, the matrix material occupies the volume fraction $(1 - f)$.

The standard constitutive variables in the pertinent literature on elastoplastic damage, e.g. [50], are the macroscopic Cauchy stress $\boldsymbol{\sigma}$ and rate of deformation $\boldsymbol{\delta}$. A rate-elastic constituent of the material with damage may be specified by the relation

$$\breve{\boldsymbol{\sigma}} = \boldsymbol{\kappa}(f)\,[\boldsymbol{\delta} - \boldsymbol{\delta}^p] \tag{167}$$

where $\breve{\boldsymbol{\sigma}}$ denotes the Jaumann co-rotational rate of $\boldsymbol{\sigma}$. The instantaneous elastic stiffness matrix $\boldsymbol{\kappa}$ of the material is assumed here to depend on the void volume fraction f which serves as a measure for the material damage. A yield condition describing the macrocospic plastic properties of the material may be stated as

$$\phi(\boldsymbol{\sigma}, \sigma_0, f) = 0 \tag{168}$$

in conformity with the usual conventions but for the appearance of the void volume fraction f, which enters as a softening constituent due to damage. The hardening variable is represented by σ_0, the uniaxial yield stress of the matrix material. Obviously, the yield condition must reduce to that for the undamaged material if $f = 0$.

During plastic flow consistency requires $\dot{\phi} = 0$, and thus from (168)

$$\frac{\partial \phi}{\partial \boldsymbol{\sigma}} \check{\boldsymbol{\sigma}} = -\frac{\partial \phi}{\partial \sigma_0} \dot{\sigma}_0 - \frac{\partial \phi}{\partial f} \dot{f} \gtreqless 0 \tag{169}$$

where a softening term appears in addition to the customary hardening term on the right-hand side of (169). As a result of plastic flow accompanied by damage, the variation of stress may point outwards, inwards, or be tangential to the yield surface $\phi = 0$ in stress space. This may be interpreted as an overall hardening, softening, or a transition, respectively.

Plastic flow is assumed to obey the normality rule also in a material with damage [47]. Therefore, the plastic part of the rate of deformation reads

$$\boldsymbol{\delta}^p = \lambda \left[\frac{\partial \phi}{\partial \boldsymbol{\sigma}} \right]^t \tag{170}$$

and is directed along the exterior normal to the yield surface in stress space at the instant of yield, its intensity being indicated by the positive scalar λ. With reference to the flow rule (170) the second-order work term

$$\check{\boldsymbol{\sigma}}^t \boldsymbol{\delta}^p \gtreqless 0 \tag{171}$$

may be seen to conform in sign with the variations of the yield function with stress on the left hand side in (169).

Hardening and Softening Constituents

Hardening of the matrix material is described by the relation

$$\dot{\sigma}_0 = h_0 \delta_0^p = h_0 \bar{\delta}_0^p = \frac{h_0}{(1-f)\sigma_0} \boldsymbol{\sigma}^t \boldsymbol{\delta}^p \tag{172}$$

where h_0 denotes the uniaxial hardening coefficient. A transition from the uniaxial state (σ_0, δ_0^p) to multiaxial conditions is performed via the equivalent quantities $\bar{\sigma}_0, \bar{\delta}_0^p$ when the matrix is assumed to undergo isochoric plastic flow obeying the von Mises yield condition. The final expression in (172) in terms of the macroscopic variables $\boldsymbol{\sigma}, \boldsymbol{\delta}^p$ relies on the rate of work relation

$$\boldsymbol{\sigma}^t \boldsymbol{\delta}^p = (1-f)\boldsymbol{\sigma}_0^t \boldsymbol{\delta}_0^p = (1-f)\bar{\sigma}_0 \bar{\delta}_0^p \tag{173}$$

which states that the rate of plastic work appearing macroscopically is exclusively a result of the plastic flow in the matrix occupying the volume fraction $(1-f)$.

In the present context damage is attributed to the nucleation and growth of voids in the material. Accordingly, the time rate of the void volume fraction may be composed as

$$\dot{f} = \dot{f}_n + \dot{f}_g \tag{174}$$

In compliance with a plastically isochoric matrix, the volumetric part of the macroscopic plastic rate of deformation is solely due to the growth of voids in the material. Therefore, one derives the growth component as

$$\dot{f}_g = (1-f)\boldsymbol{e}^t \boldsymbol{\delta}^p \tag{175}$$

where $\boldsymbol{e} = \{111000\}$ is the summation vector.

Void nucleation may be strain and/or stress controlled, cf. [48, 50] and is stated in the following in general terms as

$$\dot{f}_n = \boldsymbol{\alpha}^t \boldsymbol{\delta}^p + \boldsymbol{\beta}^t \breve{\boldsymbol{\sigma}} \tag{176}$$

the vectors $\boldsymbol{\alpha}, \boldsymbol{\beta}$ comprising the appropriate moduli. The appearance of the rate $\dot{f}_n$ (176) is conditional, but not necessarily on plastic yield.

Stress–Strain Relations

Considering the consistency condition (169) in conjunction with (172), (174) to (176) and the flow rule (170), the positive scalar λ is derived as

$$\lambda = \frac{1}{h}\left[\frac{\partial\phi}{\partial\boldsymbol{\sigma}} + \frac{\partial\phi}{\partial f}\boldsymbol{\beta}^t\right]\breve{\boldsymbol{\sigma}} = \frac{1}{h^*}\left[\frac{\partial\phi}{\partial\boldsymbol{\sigma}} + \frac{\partial\phi}{\partial f}\boldsymbol{\beta}^t\right]\boldsymbol{\kappa}\boldsymbol{\delta} \geq 0 \tag{177}$$

The second expression in (177) implies the use of the rate-elastic relation (167). Besides, we have introduced the abbreviations

$$h = -\frac{\partial\phi}{\partial\boldsymbol{\sigma}}\left[\frac{\partial\phi}{\partial\sigma_0}\frac{h_0}{(1-f)\sigma_0}\boldsymbol{\sigma} + \frac{\partial\phi}{\partial f}(1-f)\boldsymbol{e} + \frac{\partial\phi}{\partial f}\boldsymbol{\alpha}\right] \tag{178}$$

and

$$h^* = h + \left[\frac{\partial\phi}{\partial\boldsymbol{\sigma}} + \frac{\partial\phi}{\partial f}\boldsymbol{\beta}^{\mathrm{t}}\right]\boldsymbol{\kappa}\left[\frac{\partial\phi}{\partial\boldsymbol{\sigma}}\right]^{\mathrm{t}} \tag{179}$$

The plastic part of the rate of deformation is then obtained with (170) and, in conformity with (177), it may be expressed either in terms of the stress rate or in terms of the rate of deformation. Substitution of the result in the rate-elastic relation (167) yields the ultimate stress–strain relationship,

$$\check{\boldsymbol{\sigma}} = \left\{\boldsymbol{\kappa} - \frac{1}{h^*}\boldsymbol{\kappa}\left[\frac{\partial\phi}{\partial\boldsymbol{\sigma}}\right]^{\mathrm{t}}\left[\frac{\partial\phi}{\partial\boldsymbol{\sigma}} + \frac{\partial\phi}{\partial f}\boldsymbol{\beta}^{\mathrm{t}}\right]\boldsymbol{\kappa}\right\}\boldsymbol{\delta} \tag{180}$$

and the inverse relation reads

$$\boldsymbol{\delta} = \left\{\boldsymbol{\kappa}^{-1} + \frac{1}{h}\left[\frac{\partial\phi}{\partial\boldsymbol{\sigma}}\right]^{\mathrm{t}}\left[\frac{\partial\phi}{\partial\boldsymbol{\sigma}} + \frac{\partial\phi}{\partial f}\boldsymbol{\beta}^{\mathrm{t}}\right]\right\}\check{\boldsymbol{\sigma}} \tag{181}$$

With reference to (177), the parameter h defined by (178) reflects the macroscopic hardening/softening behaviour of the material. Initially $h > 0$, and in order to induce plastic flow the variation of stress must increase the yield function. When the occurence and evolution of damage leads to $h < 0$, plastic flow enforces softening, i.e. variations of stress towards a decreasing yield function. Such a variation of stress may alternatively be indicative of elastic unloading.

Despite the difference in the definitions, the significance of h appears as in standard elastoplasticity. Accordingly, the description of plastic flow may be based on the stress rate $\check{\boldsymbol{\sigma}}$ as long as $h > 0$ in the first expression in (177). The second expression in (177) describes plastic flow in terms of the rate of deformation $\boldsymbol{\delta}$ and is applicable whenever $h^* > 0$ and the rate quantity $\boldsymbol{\kappa\delta}$ points outwards the yield function.

Specification of the Approach

In the following, the rate-elastic relation (167) is assumed to depend on damage via the elastic modulus

$$E(f) = (1 - f)E_0 \tag{182}$$

E_0 referring to the original material with $f = 0$. Furthermore, a particular form of the yield condition (168) is borrowed from the modelling of plastic flow in porous solids in [53] and is generalised here as

$$\phi = \left(\frac{\bar{\sigma}}{a}\right)^2 + \left(\frac{3\sigma_H}{2b}\right)^2 - \sigma_0^2 = 0 \tag{183}$$

In the yield condition (183) $\bar{\sigma}$ and σ_H denote the macroscopic von Mises equivalent stress and hydrostatic stress respectively; the yield stress of the matrix material is σ_0. Damage is implied in the yield condition by the coefficients

$$a = a(f) \quad ; \quad 1 \geq a \geq 0 \tag{184}$$

and

$$b = b(f) \quad ; \quad \infty \geq b \geq 0 \tag{185}$$

The functional dependence of the coefficients a, b on the void volume fraction f is characteristic of damage and failure of the particular material, cf. [51, 52].

By virtue of the yield condition (183) plastic flow as given by (170) appears macroscopically to consist of a deviatoric part

$$\boldsymbol{\delta}_D^p = \lambda \frac{2}{a}\left(\frac{\bar{\sigma}}{a}\right)\boldsymbol{s} \quad , \quad \boldsymbol{s} = \frac{3}{2}\frac{1}{\bar{\sigma}}\boldsymbol{\sigma}_D \tag{186}$$

and a volumetric part

$$\boldsymbol{\delta}_V^p = \lambda \frac{1}{b}\left(\frac{3\sigma_H}{2b}\right)\boldsymbol{e} \tag{187}$$

When considering the evolution of damage by (174), the growth of voids (175) reads

$$\dot{f}_g = (1-f)\frac{3\lambda}{b}\left(\frac{3\sigma_H}{2b}\right) \tag{188}$$

Void nucleation may be stated in the form, cf. [48, 50],

$$\dot{f}_n = A\dot{\sigma}_0 + B\dot{\sigma}_H \tag{189}$$

in which case the moduli in (176) read

$$\boldsymbol{\alpha} = \frac{h_0 A}{(1-f)\sigma_0}\boldsymbol{\sigma} \tag{190}$$

and

$$\beta = \frac{B}{3} e \tag{191}$$

The material functions A, B may be taken as suggested in [50]. However, further investigations appear necessary in this respect, cf. also [52].

Applications

A first application of the constitutive relations with damage concerns the extension of the axisymmetric cylindrical specimen in Fig. 39. The characterisation of the steel material relies on the data reported in [54, 55]. The uniaxial true stress-logarithmic strain diagram depicted in Fig. 39 refers to the material without damage and reflects the hardening properties of the matrix. Inclusions participate with a volume fraction $f_0 = 1.6 \times 10^{-4}$ in the material and are considered to completely convert into voids when the nucleation criterion

$$\sigma_1 + 1.6(\bar{\sigma} - \sigma_0) = 1120 \text{ MPa} \tag{192}$$

is met locally; σ_1 denotes the maximum principal stress, σ_0 is taken as the initial yield stress. By an interpretation of the investigations in [54], the parameters a, b in the present yield conditions (183) are specified as (Damage 1)

$$a = (1 - f)(1 - 1.27f) \quad ; \quad b = -1.27(1 - f)\ln(1.27f) \tag{193}$$

For the purpose of comparison the functions

$$a = (1 - f)^2 \quad ; \quad b = -\ln f \tag{194}$$

are alternatively employed (Damage 2). In addition, a critical value $f_c = 0.15$ in the first and $f_c = 0.25$ in the second case indicates the complete vanishing of strength in the material.

Results are obtained along the lines of an ordinary elastoplastic computation [21] modified for the inclusion of damage, and are shown in Figs. 40 to 43. The onset of necking in the specimen is neither driven by an initial geometric imperfection nor does it necessitate any particular considerations in the present numerical analysis. When the material does not undergo damage, it is able to sustain extensive local deformations (cf. Fig. 42), but for this reason the results obtained for relative elongations far beyond 0.3 are not considered accurate. Damage leads to a markable drop of the axial load as demonstrated in Fig. 40. The sharp knees in the diagram indicate the onset of local failure in the specimen where $f = f_c$. The introduction of this local failure criterion is seen to facilitate the computation. But for the different evolution of the damage variable f, both assumptions for $a(f), b(f)$ provide similar overall results. Because of the particular nucleation criterion (192), damage does not appear here before the onset of necking and has little influence on it (Fig. 41). Damage and failure of the material limit the deformability of the specimen as shown in Fig. 41 and Fig. 43. The evolution of failure at the neck is indicated and may be followed in Fig. 43. Also, the computational results are in gross agreement with the experimental failure behaviour reported in [55]. For an assessment of the present investigation it is important to note that the solution for the non-damaging material obtained with finite element discretisation documented in Fig. 39, compares well with that of section 4.4 based on the adaptively regenerated mesh.

The subsequent investigation concerns the extension of the axisymmetric notched specimen in Fig. 44 which is from the same material as before. This example was analysed in [54] on a different constitutive basis and is computed here along the lines of the foregoing investigation. Results of the present analysis with a, b as defined by (193), but with $f_c = 0.125$ are depicted in Figs. 45 and 46. The variation of the axial load with the contraction of the notch in Fig. 45 is in agreement with the experimental results reported in [54]. The alternative approach (194) to a, b leads here to an unrealistic decrease of the load. In Fig. 46, the deformation of the notch is shown for an extension $l - {}^0 l = 0.58$ mm of the specimen at which incipient failure has taken place in the center of the specimen as indicated in the figure.

The further studies concern flat specimens from St37 steel material. In this case, nucleation of voids is supposed to be completed within the range of homogeneous deformation in the tensile specimen in Fig. 47. Therefore, when the hardening properties of the matrix are to be deduced from the tension test, the influence of damage has to be eliminated (cf. Fig. 47). An estimate of the uniaxial characteristic of the material without damage beyond initial yield may be given by the relation

$$\sigma = 520\gamma^{0.25} \text{ MPa} \tag{195}$$

between the true stress and logarithmic strain. Also, for this material $E = 210000$ MPa, $\nu = 0.3$. For damage, a, b are taken as by (194), but in the present case $f_c = 1$ is used in the computation. The volume fraction of inclusions amounts to approximately $f_0 = 0.5 \times 10^{-2}$. This is assumed for simplicity to provide an equal initial void volume fraction at a logarithmic plastic strain $\eta = 0.05$. The effect of this assumption and of the subsequent evolution of damage is clearly shown in the load-elongation diagram in Fig. 48. In this figure also the good agreement with experimental data is demonstrated. The influence of damage on the onset of necking in the present case is apparent from Fig. 49.

As a last example consider the flat sheet with a circular hole in Fig. 50. The material is again St37 steel for which, however, the range of the experimental data available in Fig. 47 is limited. Therefore, the material cannot be uniquely characterised when large strains occur in the present specimen and the previous model is not immediately applicable. Instead, a first analysis is based on an unmodified elastoplastic computation.

In this study the possibility is explored to account for the effects of damage directly via the uniaxial macroscopic hardening properties of the material. For this purpose, the yield stress is specified as a function of plastic strain in the range of the expected large deformations, such that computation reproduces the raising branch of the experimental load-extension diagram of Fig. 51. Continuation of the computation beyond that stage reveals in fact the maximum load and a postcritical behaviour of the specimen not much too different from the experimental test. This remark applies also to the deformations shown in Fig. 52 for the experiment, [56], and in Fig. 53 for the computation.

Acknowledgement

The author wishes to express his gratitude to the Director of the Institute for Computer Applications, Professor J. Argyris, F.R.S., for the generous support provided for the preparation of the paper. The layout and typing of the manuscript by Ms J. Cuenin is once more gratefully acknowledged.

References

1 J. Argyris, H.-P. Mlejnek, Die Methode der Finiten Elemente, Band I, II, III (Friedrich Vieweg u. Sohn, Braunschweig Wiesbaden, 1987).

2 W. Prager, Introduction to Mechanics of Continua (Ginn, Boston,1961).

3 O.C. Zienkiewicz, The Finite Element Method (McGraw Hill, London, 1977).

4 J. Argyris, K. Straub and Sp. Symeonidis, Static and dynamic stability of nonlinear elastic systems under non-conservative forces – natural approach, Comput. Meths. Appl. Mech. Engrg. 32 (1982) 59-83.

5 J. Argyris, Continua and discontinua, Matrix Methods in Structural Mechanics, Opening address, Proceedings of the Conference held at Wright-Patterson Air Force Base, Ohio, 26-28 Oct. 1965.

6 E. Riks, An incremental approach to the solution of snapping and buckling problems, Int. J. Sol. Struct. 15 (1979) 529-551.

7 J. Argyris, H. Balmer, I.St. Doltsinis, Implantation of a nonlinear capability on a linear software system, Comput. Meths. Appl. Mech. Engrg. 65 (1987) 267-291.

8 J.-L. Batoz and G. Dhatt, Incremental displacement algorithms for nonlinear problems, Int. J. Num. Meths. Eng. 14 (1979) 1262-1267.

9 M.A. Crisfield, A fast incremental/iterative solution procedure that handles "snap-through", Computers & Structures 13 (1981) 55-62.

10 E. Ramm, Strategies for tracing the nonlinear response near limit points, in: W. Wunderlich, E. Stein and K.J. Bathe, eds., Nonlinear Finite Element Analysis in Structural Mechanics (Springer, Berlin, 1981) 63-89.

11 I. Fried, Orthogonal trajectory accession to the nonlinear equilibrium curve, Comput. Meths. Appl. Mech. Engrg. 47 (1984) 283-297.

12 J. Argyris and D.W. Scharpf, Berechnung vorgespannter Netzwerke, Bayerische Akademie der Wissenschaften, Sonderdruck 4 (1970) 25-58.

13 J. Argyris and P.C. Dunne, A simple theory of geometrical stiffness with applications to beam and shell problems, Second International Symposium on Computing Methods in Applied Sciences and Engineering, Versailles, France, Dec. 15-19, 1975, ISD Report No. 183, Stuttgart, 1975.

14 H.A. Balmer and I.St. Doltsinis, An account of the geometrically nonlinear analysis of structures with ASKA, ICA Report for Saab-Scania, Stuttgart, 1986.

15 J. Argyris, P.C. Dunne, G.A. Malejannakis and E. Schelkle, A simple triangular facet shell element with applications to linear and nonlinear equilibrium and elastic stability problems, Comput. Meths. Appl. Mech. Engrg. 10 (1977) 371-403.

16 J. Argyris, An excursion into large rotations, Comput. Meths. Appl. Mech. Engrg. 32 (1982) 85-155.

17 R. Hill, Some basic principles in the mechanics of solids without a natural time, J. Mech. Phys. Solids 7 (1959) 209-225.

18 E.H. Lee, Elasto-plastic deformation at finite strain, J. Appl. Mech. 36 (1961) 1-6.
19 A.E. Green and P.M. Naghdi, A general theory of an elastic-plastic continuum, Arch. Rat. Mech. Ann. 18 (1965) 251-281.
20 J. Argyris and I.St. Doltsinis, On the large strain inelastic analysis in natural formulation. Part I. Quasistatic problems, Comput. Meths. Appl. Mech. Engrg. 20 (1979) 213-251; Part II. Dynamic Problems, Comput. Meths. Appl. Mech. Engrg. 21 (1980) 91-128.
21 J. Argyris, I.St. Doltsinis, P.M. Pimenta and H. Wüstenberg, Thermomechanical response of solids at high strains – natural approach, Comput. Meths. Appl. Mech. Engrg. 32 (1982) 3-57.
22 H. Balmer, I.St. Doltsinis and M. König, Elastoplastic and creep analysis with the ASKA program system, Comput. Meths. Appl. Mech. Engrg. 3 (1974) 87-104.
23 J. Argyris and I.St. Doltsinis, On the integration of inelastic stress-strain relations. Part 1: Foundations of method; Part 2: Developments of method, Res Mech. Lett. 1 (1981) 343-355.
24 T.J.R. Hughes, Numerical implementation of constitutive models: Rate-independent deviatoric plasticity, in: S. Nemat-Nasser et al., eds., Theoretical Foundation of Large-Scale Computations of Nonlinear Material Behavior (Martinus Nijhoff Publishers, Dordrecht, 1984).
25 J. Argyris, I.St. Doltsinis and M. Kleiber, Incremental formulation in nonlinear mechanics and large strain elasto-plasticity – natural approach. Part II, Comput. Meths. Appl. Mech. Engrg. 14 (1978) 259-294.
26 J. Argyris and I.St. Doltsinis, A primer on superplasticity in natural formulation, Comput. Meths. Appl. Mech. Engrg. 46 (1984) 83-131.
27 J. Argyris, I.St. Doltsinis, H. Fischer, H. Wüstenberg, '$T\alpha\ \pi\alpha\nu\tau\alpha\ \rho\epsilon\iota$', Comput. Meths. Appl. Mech. Engrg. 51 (1985) 289-362.
28 O.C. Zienkiewicz, Flow formulation for numerical solution of forming processes, in: J.F.T. Pittman et al., eds., Numerical Analysis of Forming Processes (Wiley, Chichester, 1984) 1-44.
29 J. Argyris, I.St. Doltsinis and J. Luginsland, Three-dimensional thermomechanical analysis of metal forming processes, Internat. Workshop on Simulation of Metal Forming Processes by the Finite Element Method, 3rd June 1985, Stuttgart, Proceedings (Springer, 1986).
30 R.L. Taylor and O.C. Zienkiewicz, Mixed finite element solution of fluid flow problems, in: R.H. Gallagher et al., eds., Finite Elements in Fluids, Vol. 4 (Wiley, New York, 1982).
31 J. Argyris and I.St. Doltsinis, Computer simulation of metal forming processes, Conf. on Numerical Methods in Industrial Forming Processes, Gothenburg, 25-29 August 1986, Proceedings.

32 I.St. Doltsinis, J. Luginsland and S. Nölting, Some developments in the numerical simulation of metal forming processes, Engineering Computations, 4 (1987) 266-280.
33 J.T. Oden and J.A.C. Martins, Models and computational methods for dynamic friction phenomena, Comput. Meths. Appl. Mech. Engrg. 52 (1985) 527-634.
34 I. Babuska et al., eds., Accuracy estimates and adaptive refinements in finite element computations (Wiley, New York, 1986).
35 N. Kikuchi, Adaptive grid-design methods for finite element analysis, Comput. Meths. Appl. Mech. Engrg. 55, (1986) 129-160.
36 O.C. Zienkiewicz, Y.C. Liu and G.C. Huang, Error estimation and adaptivity in flow formulation for forming problems, Int. J. Num. Meths. Engrg., 25 (1988) 23-42.
37 J. Argyris, H.A. Balmer, I.St. Doltsinis and A. Kurz, Computer simulation of crash phenomena, Int. J. Num. Meths. Engrg., 22 (1986) 497-519.
38 N.M. Newmark, A method of computation for structural dynamics, Proc. ASCE 85 (1959) 67-94.
39 J. Argyris, P.C. Dunne and Th. Angelopoulos, Dynamic response by large step integration, Earthquake Eng. Str. Dyn., 2 (1973) 185-203.
40 I.St. Doltsinis, Diskretisierungsmethoden für zeitabhängige Vorgänge, Workshop 'Diskretisierungen in der Kontinuumsmechanik', Physikzentrum Bad Honnef, 23-26 Sept. 1985, Proceedings.
41 J. Argyris and I.St. Doltsinis, On the natural formulation and analysis of large deformation coupled thermomechanical problems, Comput. Meths. Appl. Mech. Engrg. 25 (1981) 195-253.
42 J. Argyris, I.St. Doltsinis, F. Kneese and H. Wüstenberg, Numerik thermomechanischer Vorgänge, ICA Report No. 1, Stuttgart, 1984. Forschung im Ingenieurwesen 635 (1986).
43 K.A. Padmanabhan and G.J. Davies, Superplasticity (Springer, Berlin, 1980).
44 A.K. Ghosh and C.H. Hamilton, Mechanical behaviour and hardening characteristics of a superplastic Ti-6Al-4V alloy, Metallurgical Trans. A, 10A (1979) 699-706.
45 I.St. Doltsinis and S. Nölting, Numerische Simulation der superplastichen Umformung eines Schaltkastens, ICA Report for Messerschmitt-Bölkow-Blohm GmbH, Transport- u. Verkehrsflugzeuge, 1987.
46 J.L. Bluhm and R.J. Morrissey, Fracture of a tensile specimen, Proc. 1st Internat. Conf. Fract., 3 (1966)
47 A.L. Gurson, Continuum theory of ductile rupture by void nucleation and growth - I. Yield criteria and flow rules for porous ductile media, J. Engrg. Materials Technol. 99 (1977) 2-15.
48 A.L. Gurson, Porous rigid-plastic materials containing rigid inclusions – yield function, plastic potential and void nucleation, Fracture 1977, 2, ICF 4 (1977).

49 I.St. Doltsinis, Some considerations on elastoplastic solids with damage, 7th Int. Conf. on Comput. Meths. in Appl. Science and Engrg., Dec. 9-13 1985, Versailles, France, Proceedings (North-Holland, Amsterdam, 1987).
50 V. Tvergaard, Ductile fracture by cavity nucleation between larger voids, The Danish Center for Applied Mathematics and Mechanics, Report No. 210 (May 1981).
51 V. Tvergaard and A. Needleman, Analysis of the cup-cone fracture in a round tensile bar, The Danish Center for Applied Mathematics and Mechanics, Report No. 264 (June 1983).
52 P. Perzyna, On constitutive modelling of dissipative solids for plastic flow, instability and fracture, in: A. Sawczuk and G. Bianchi, eds., Plasticity Today – Modelling, Methods and applications (Elsevier Applied Science Publications, 1985)
53 R.J. Green, A plasticity theory for porous solids, Int. J. Mech. Sci. 14 (1972) 215-224.
54 G. Rousselier, K. Phan Ngoc and G. Mottet, Elastic-plastic constitutive relations including ductile fracture damage, 7th Internat. Conf. Struct. Mech. Reactor Technology, 1983, Paper G/F No. 1/3.
55 J.-C. Devaux, F. Mudry, G. Rousselier and A. Pineau, An experimental program for the validation of local ductile fracture criteria using axisymmetrically cracked bars and compact tension specimens, Engrg. Fract. Mech. 21 (1985) 273-283.
56 Institut für Bildsame Formgebung, RWTH Aachen, Private Communication (1985).

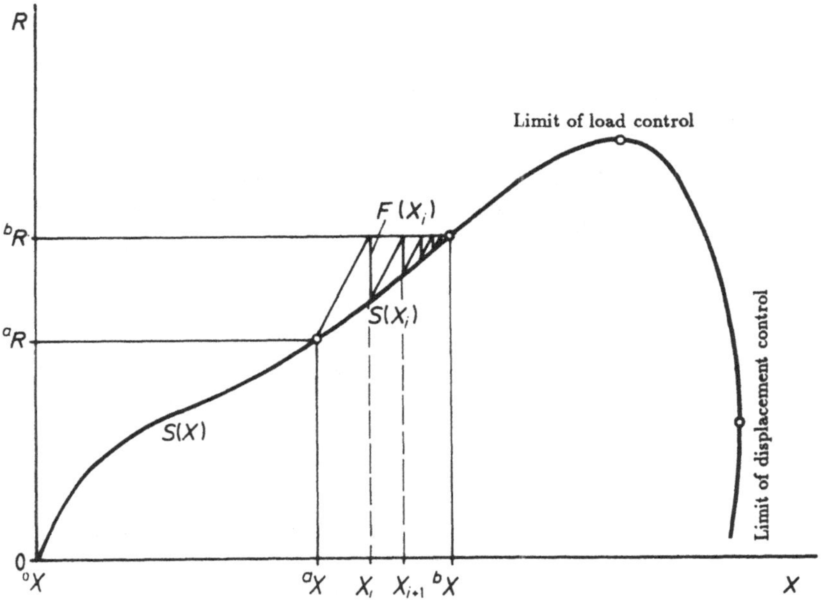

Fig. 1 Equilibrium iteration at constant load.

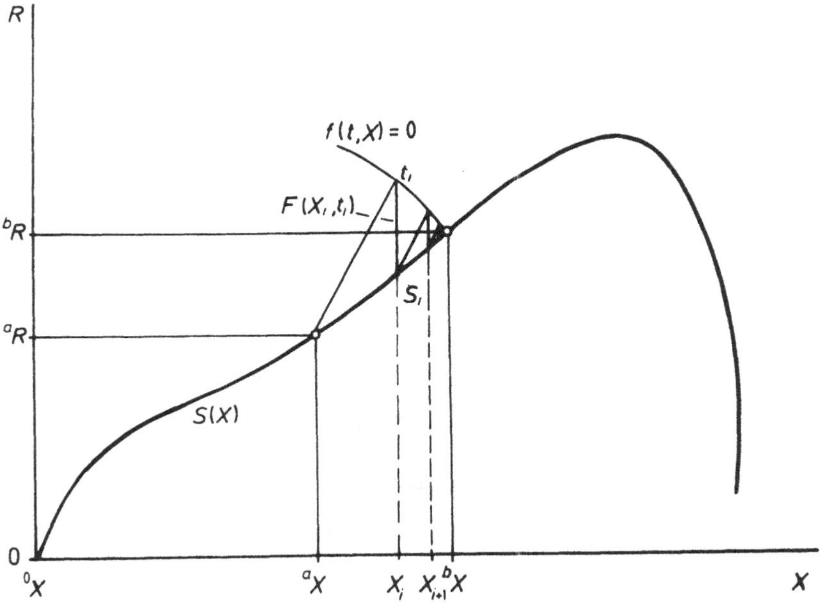

Fig. 2 Equilibrium iteration constrained by $f(t,X) = 0$.

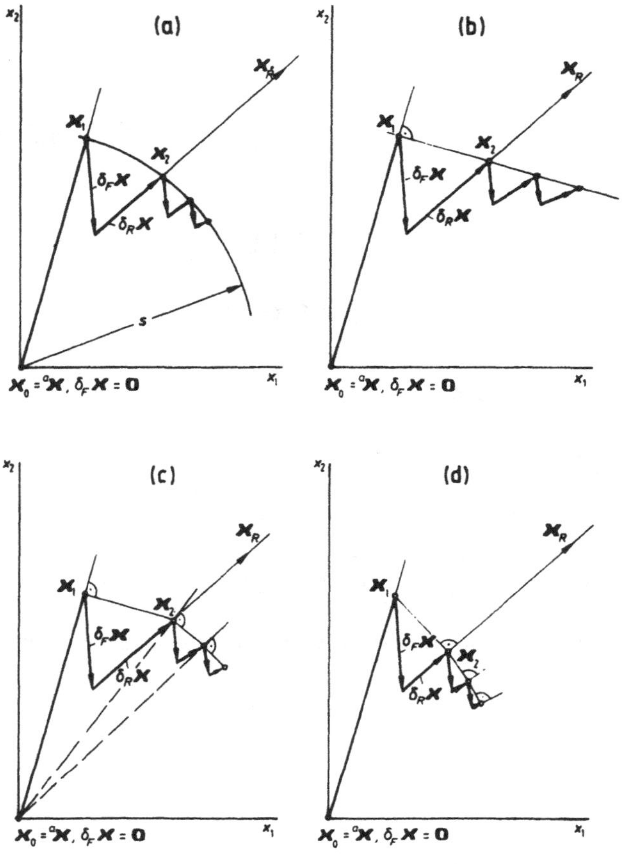

Fig. 3 Various types of the constraint.

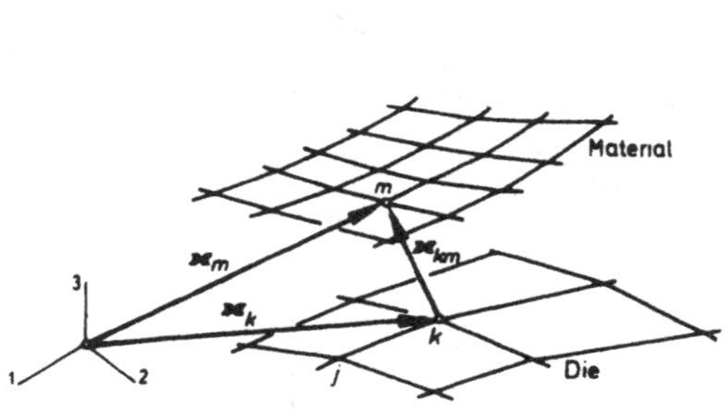

Fig. 4 Node k of die surface closest to node m of material surface.

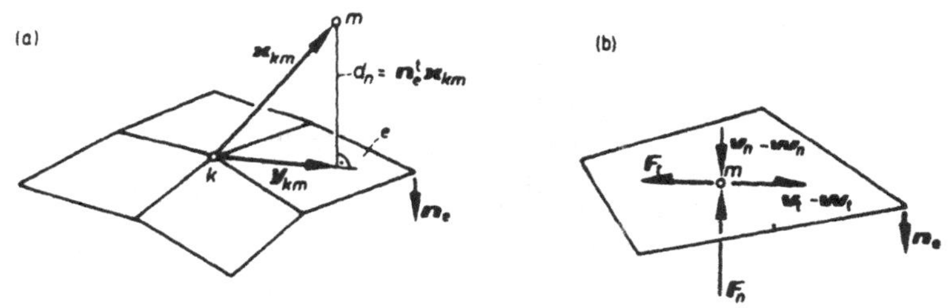

Fig. 5 (a) Plane element e closest to material node m; test for contact.
(b) Forces and velocities at contact.

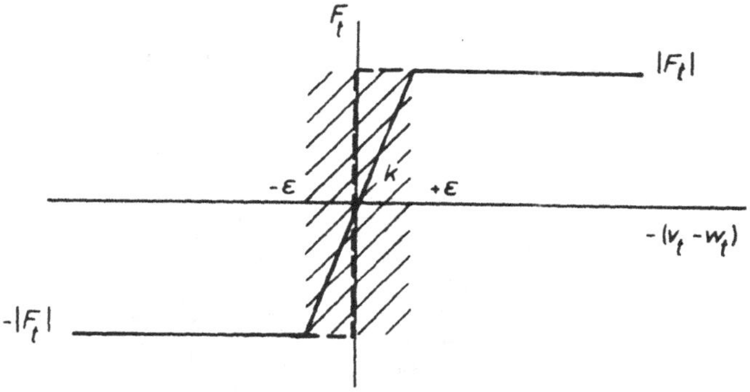

Fig. 6 Penalty approach to sticking via k.

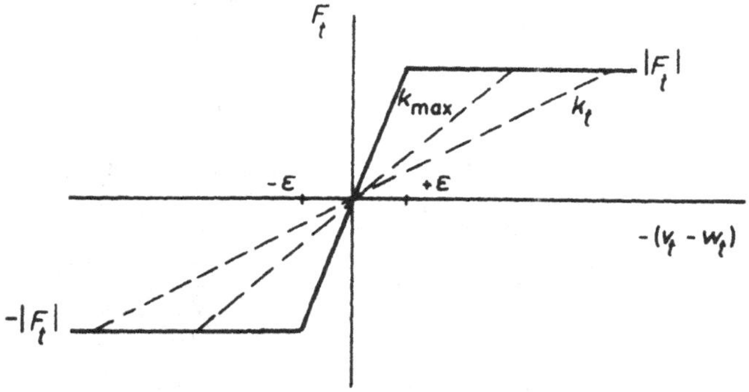

Fig. 7 Kinematic (secant) formulation of friction via k_t; limitation by $k_{\max}$ implies penalty approach to sticking.

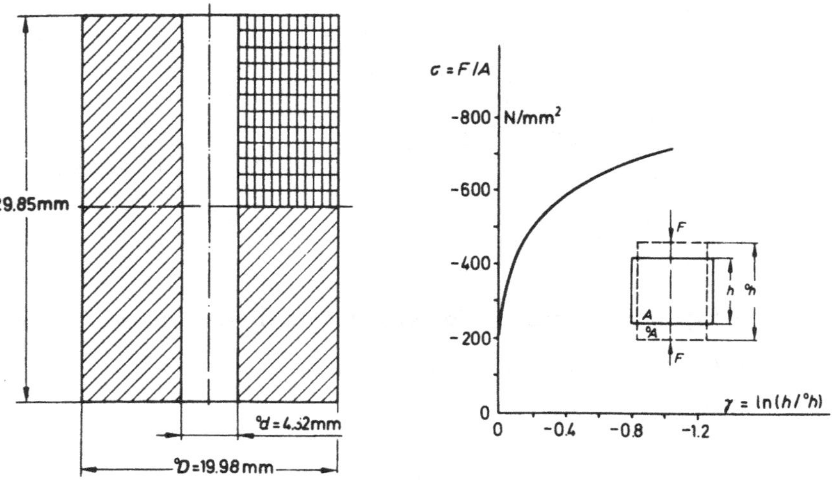

Fig. 8 Upsetting of ring specimen.

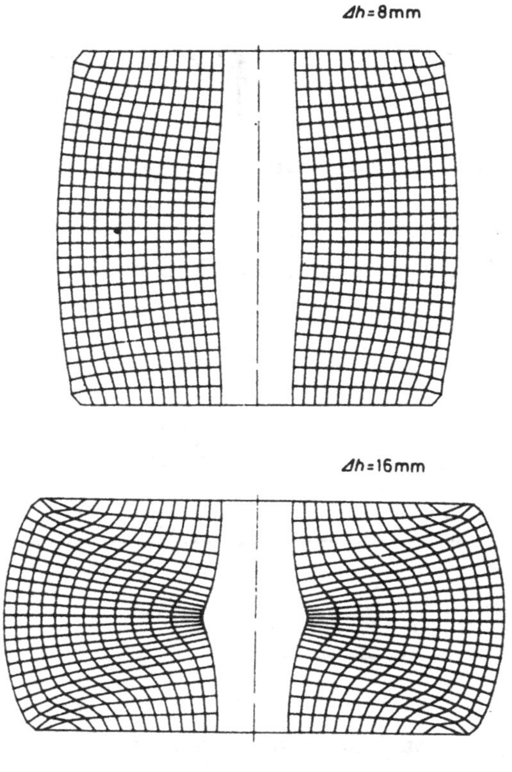

Fig. 9 Stages of deformation.

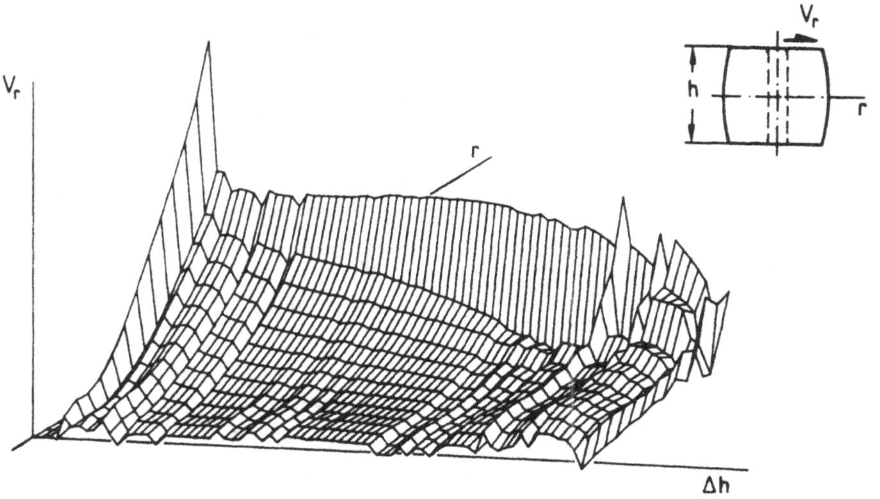

Fig. 10 Friction as external force causes oscillations of radial velocity V_r on surface of contact with die.

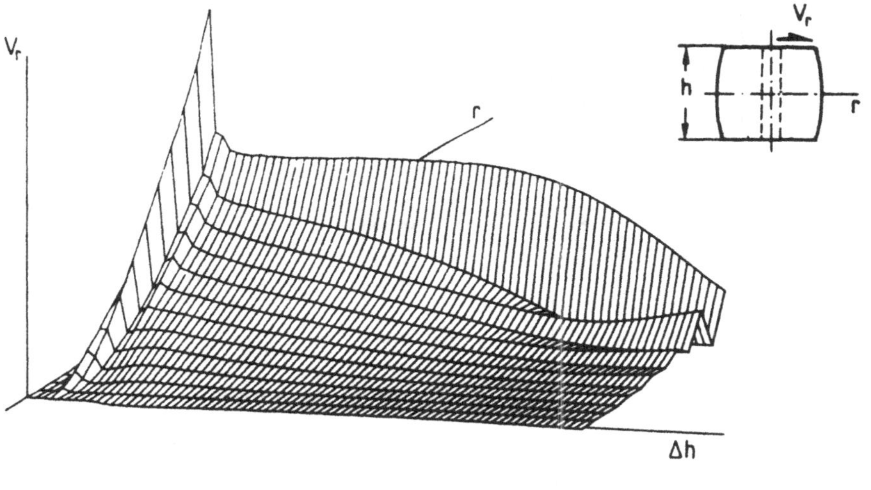

Fig. 11 Friction contributes to the system's matrix; smooth variation of radial velocity V_r on surface of contact with die.

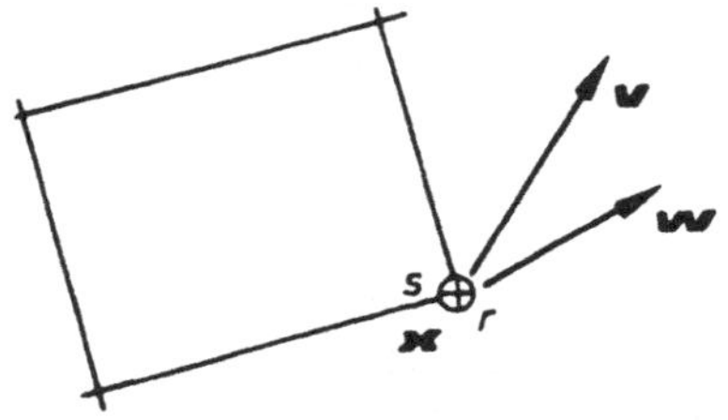

Fig. 12 Eulerian - Lagrangean technique. Mesh nodal point r moves with velocity $\boldsymbol{w}$, material point s moves with velocity $\boldsymbol{v}$.

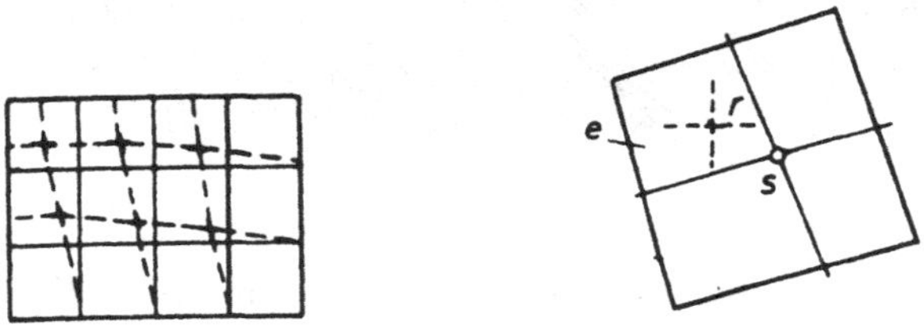

Fig. 13 Interpolation technique. Location of new nodal point r in the existing mesh (nodal point s, element e).

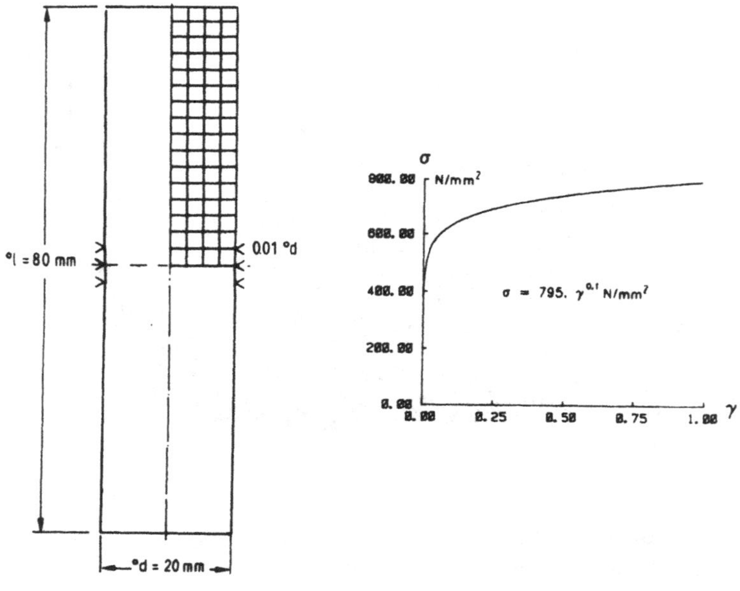

Fig. 14 Cylindrical specimen under tension.

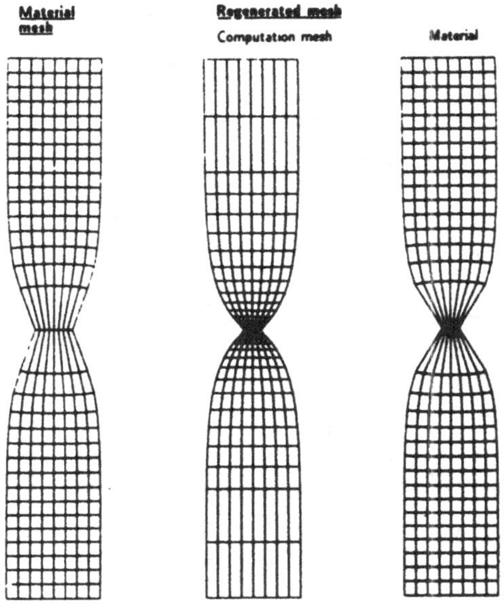

Fig. 15 Deformations for material mesh and for adaptively regenerated mesh at $(l - {}^0l)/{}^0l = 0.35$.

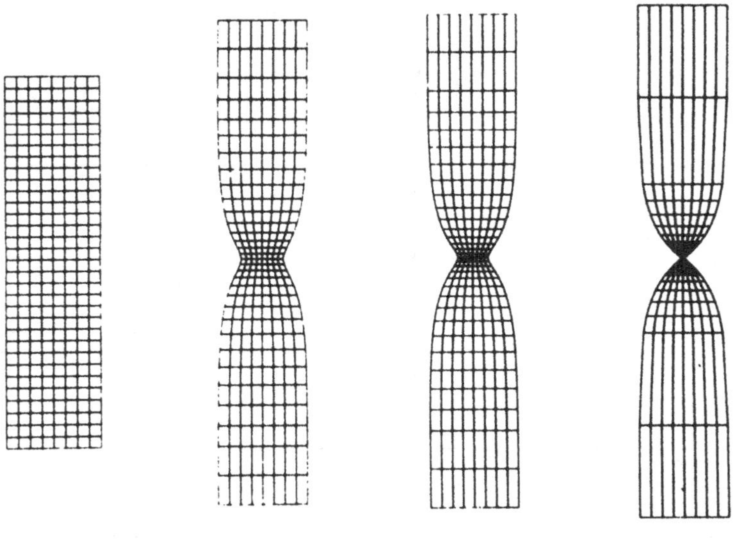

Fig. 16 Concentration of the adaptive mesh in the region of necking.

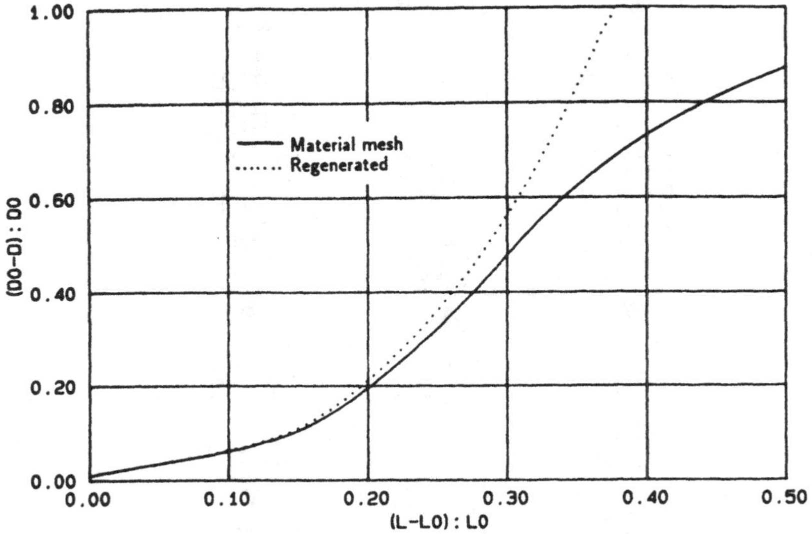

Fig. 17 Development of necking for material mesh and for adaptively regenerated mesh.

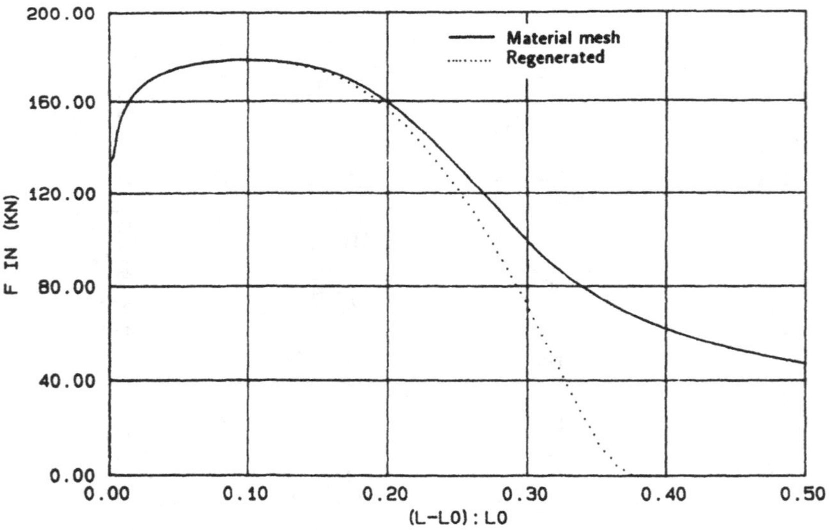

Fig. 18 Variation of the tensile force for material mesh and for adaptively regenerated mesh.

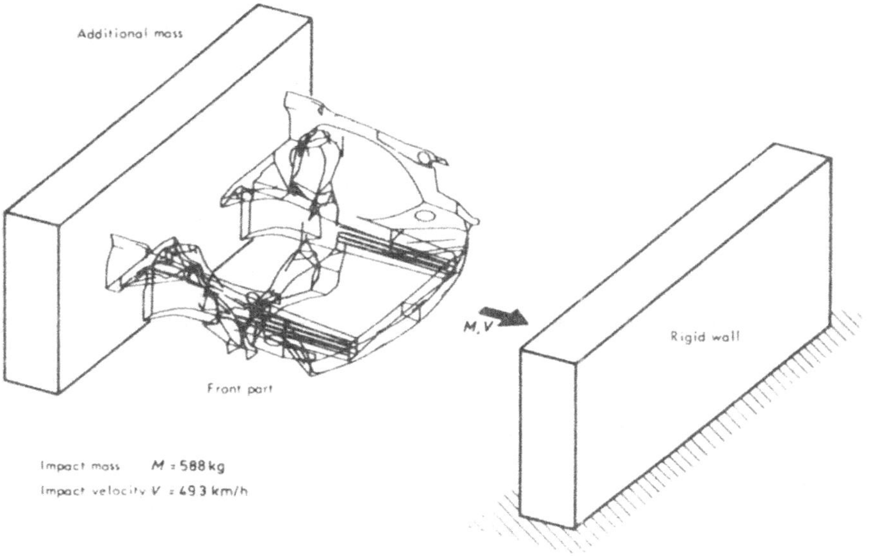

Fig. 19 Model crash of front end of car.

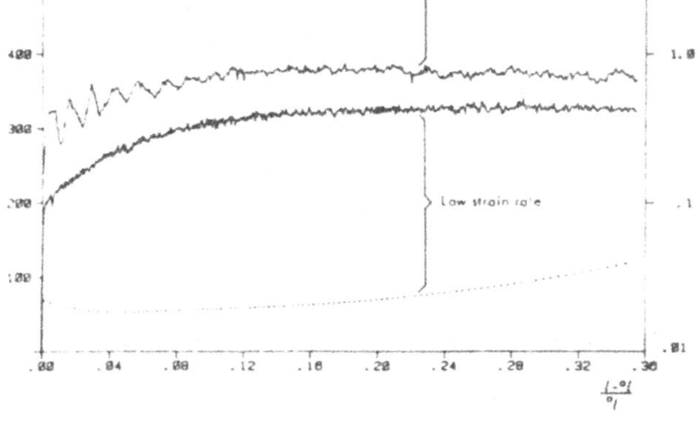

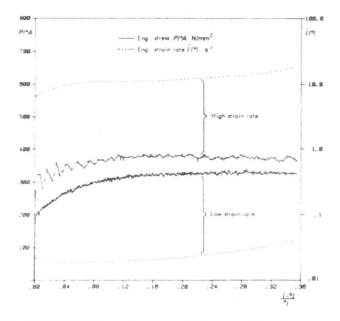

Fig. 20 Uniaxial tension test, influence of strain rate.

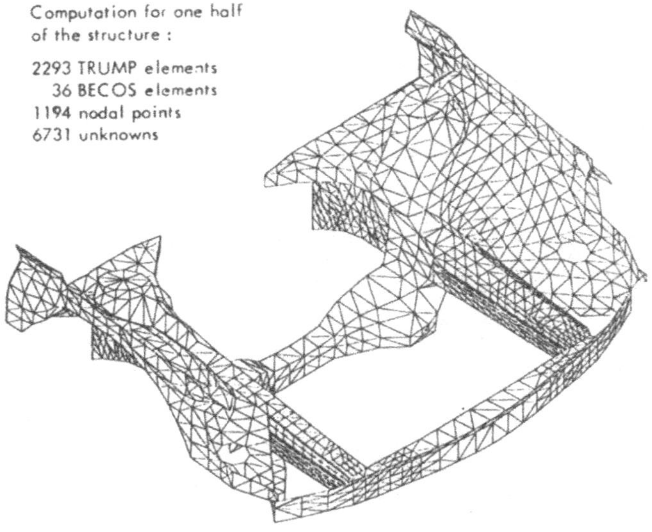

Fig. 21 Finite element model of front end of car.

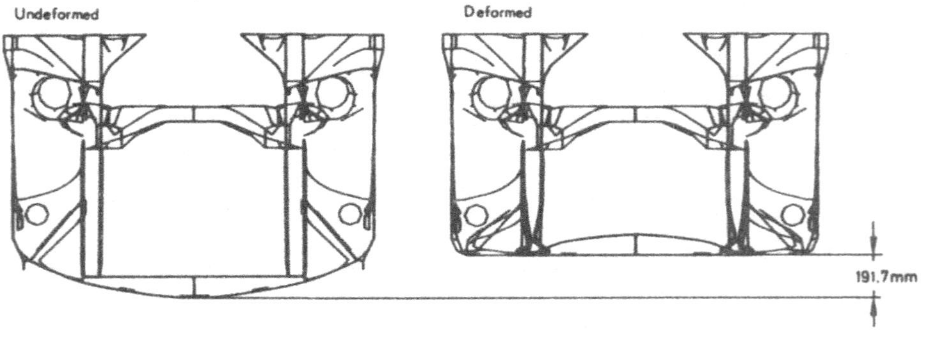

Fig. 22 Deformed stage, view from above.

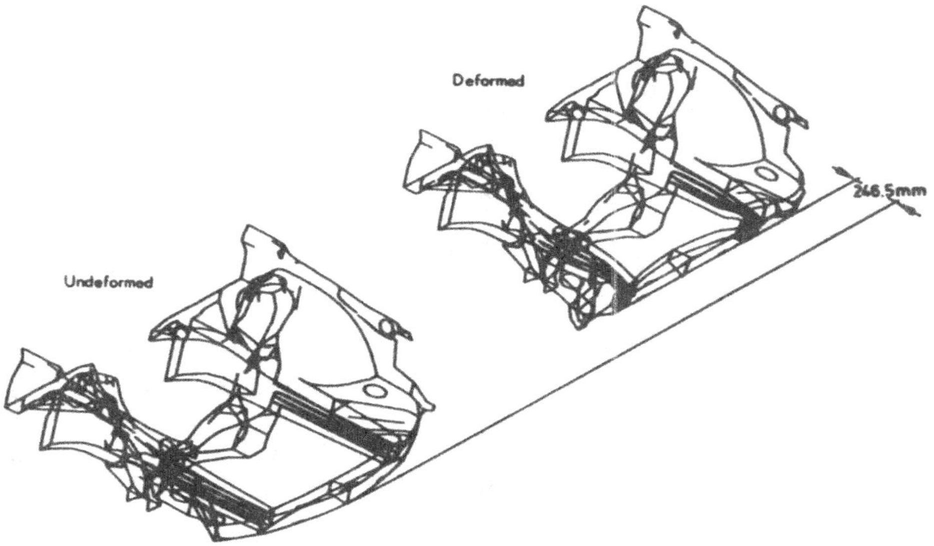

Fig. 23 Deformed stage, perspective view.

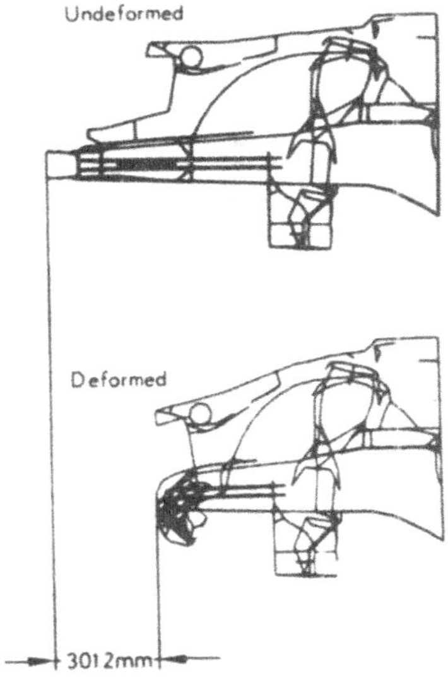

Fig. 24 Deformed stage, side view.

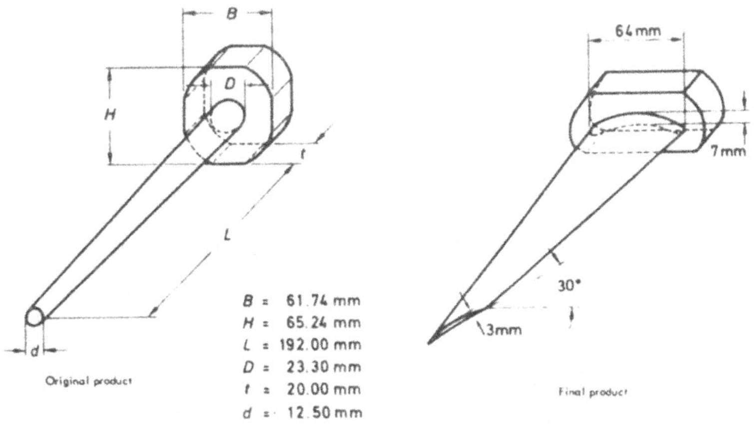

Fig. 25 Forging of a compressor blade.

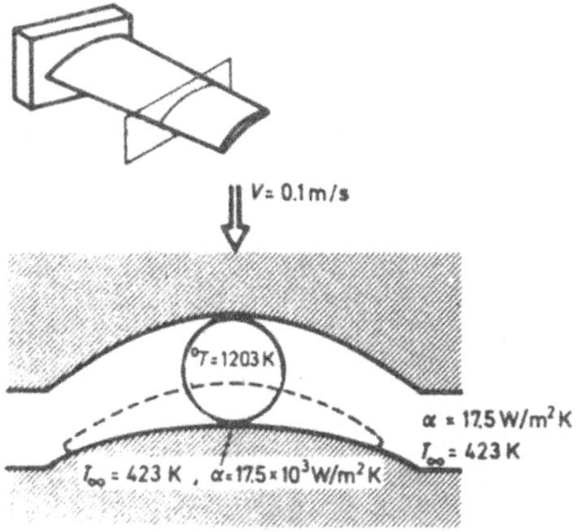

Fig. 26 Description of forging process.

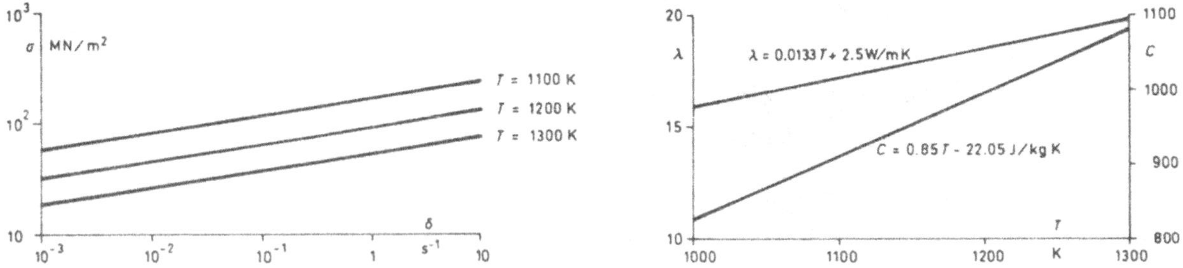

Fig. 27 Thermomechanical properties of Ti-6Al-4V material.

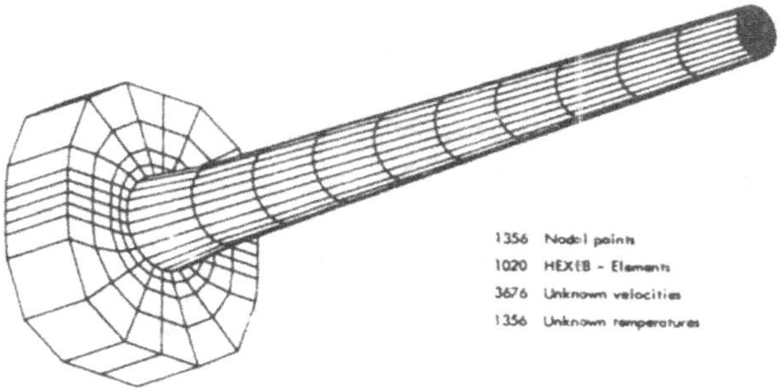

Fig. 28 Finite element mesh.

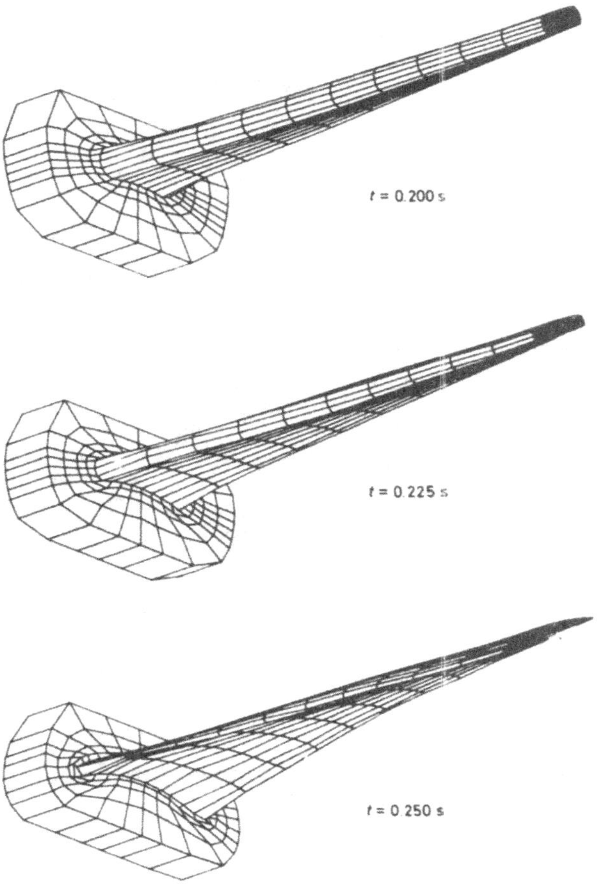

Fig. 29 Stages of deformation.

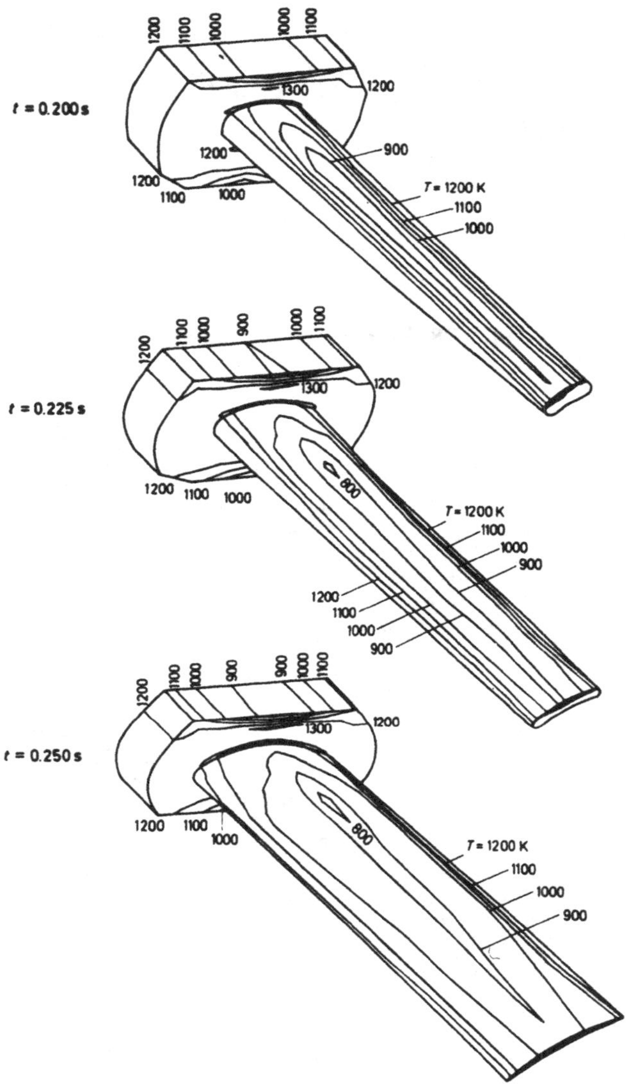

Fig. 30 Development of temperature field.

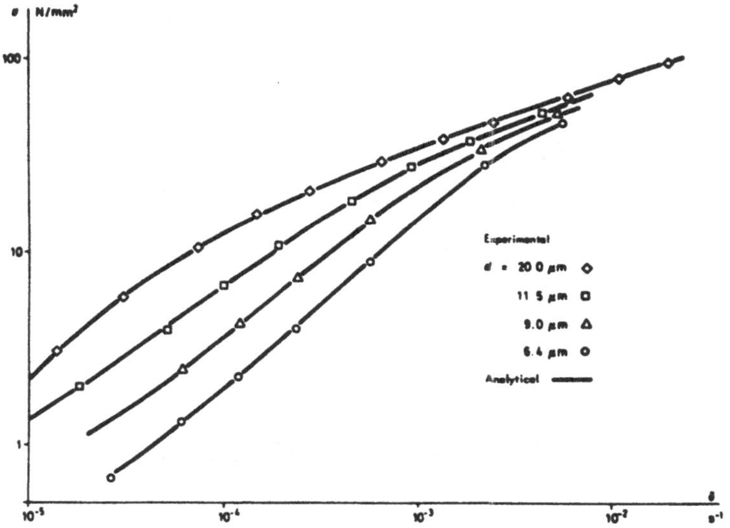

Fig. 31 Dependence of flow stress σ on rate of deformation δ in the superplastic range (Ti-6Al-4V, T=1200 K).

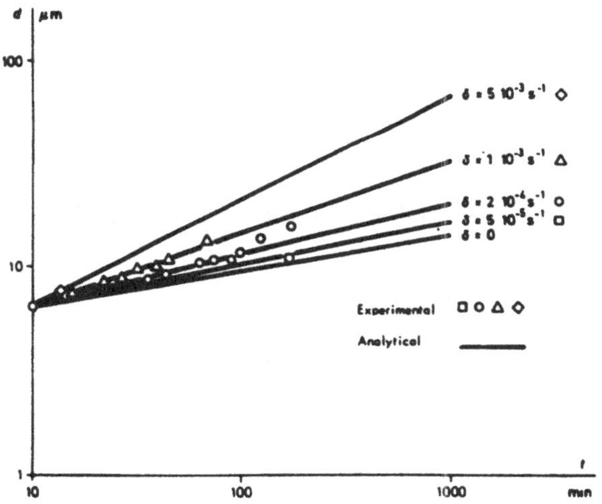

Fig. 32 Grain-growth kinetics (Ti-6Al-4V, ${}^0d$ = 6.4 μm, T=1200 K).

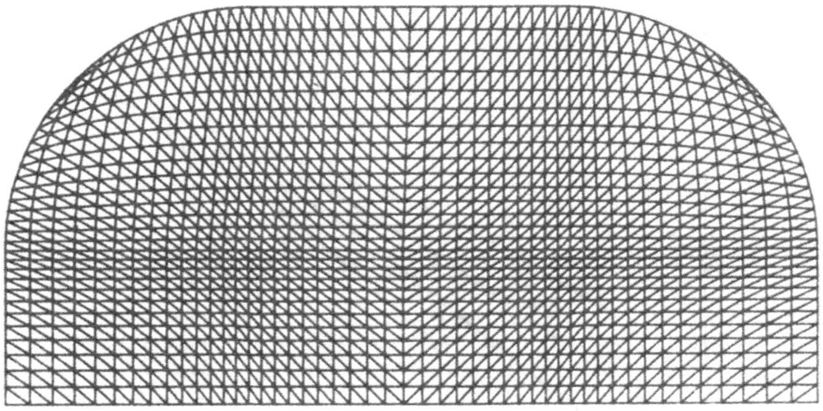

Fig. 33 Discretisation of plane sheet.

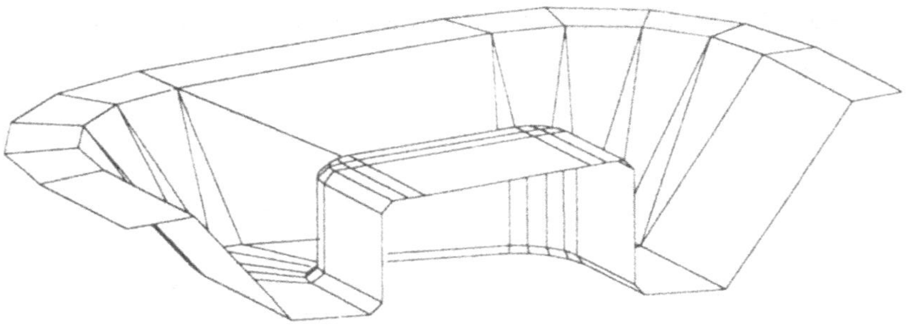

Fig. 34 Discretisation of three-dimensional die surface.

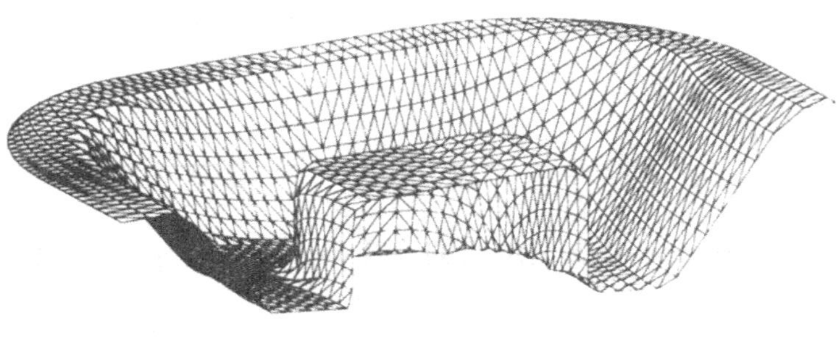

Fig. 35 Deformed sheet at $t = 6410$ s process time.

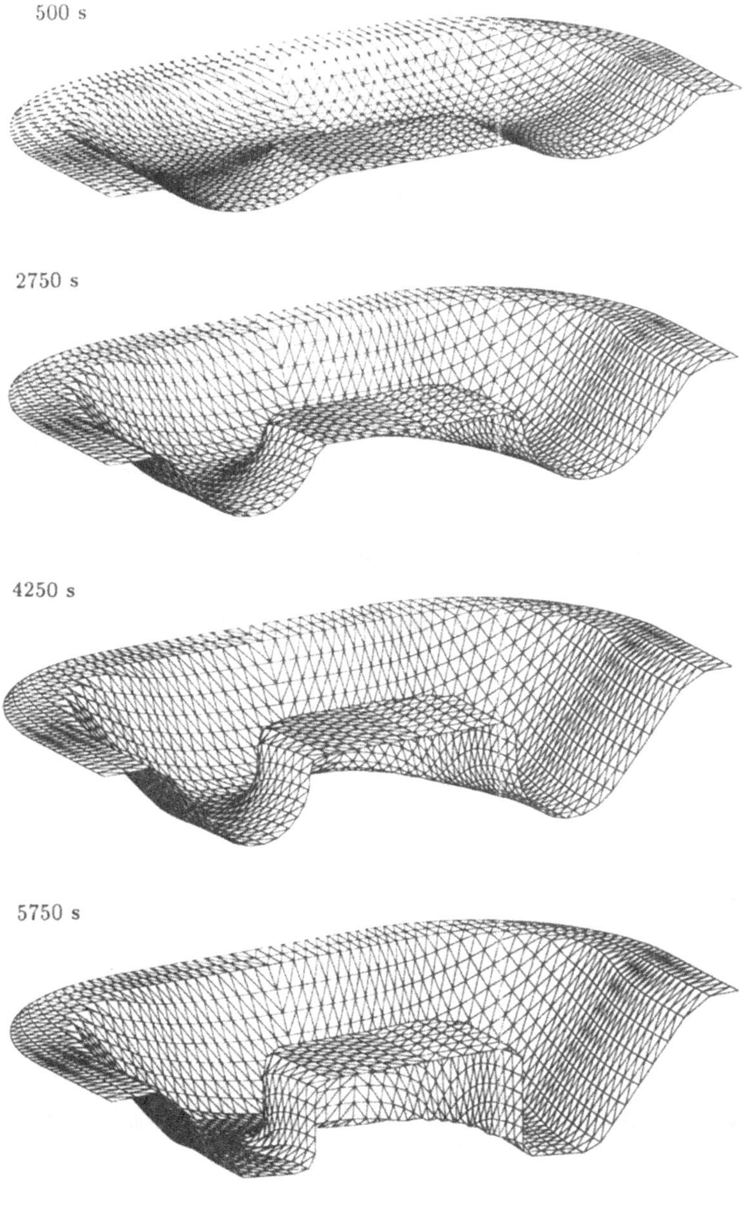

Fig. 36 Progress of the superplastic forming process.

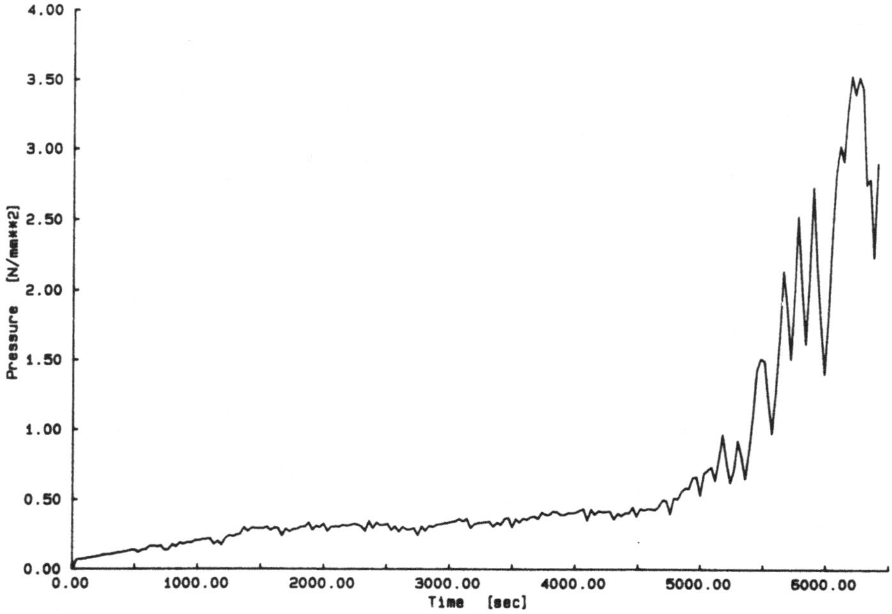

Fig. 37 Forming pressure as a function of time.

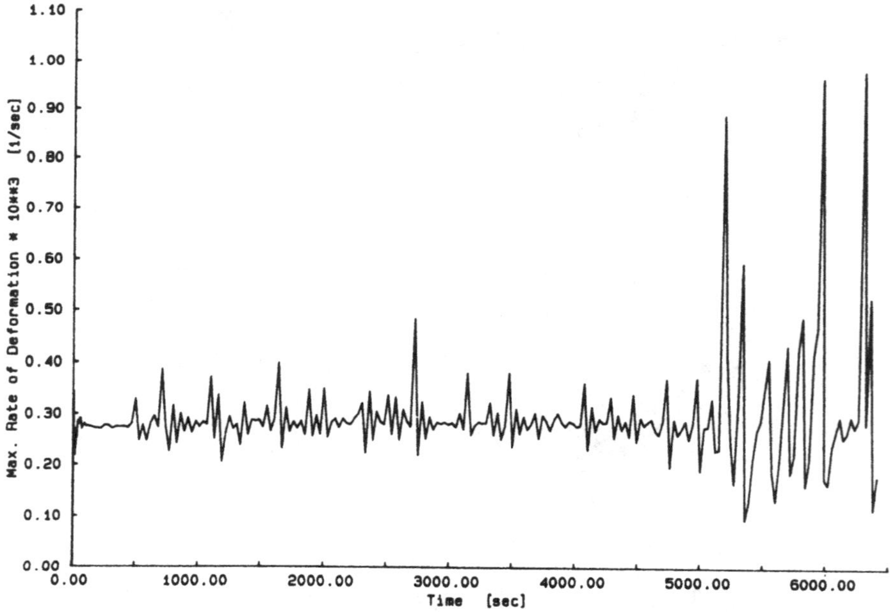

Fig. 38 Maximum rate of deformation as a function of time.

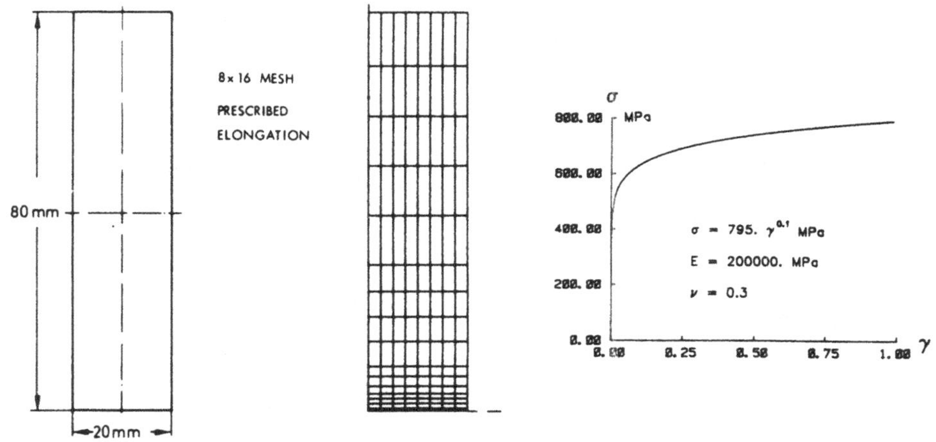

Fig. 39 Investigation of material damage. Axisymmetric tensile specimen.

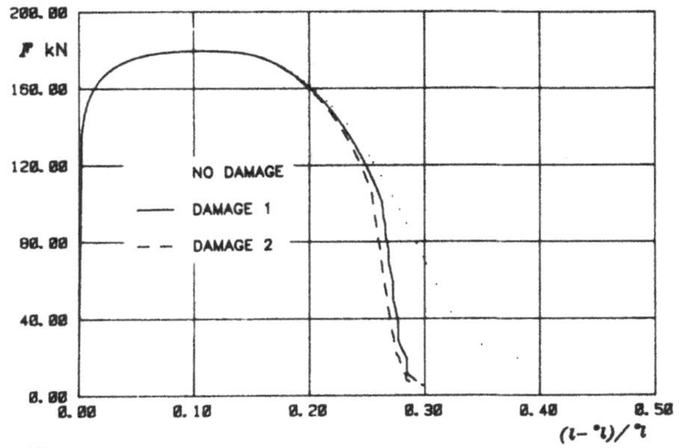

Fig. 40 Load – elongation diagram.

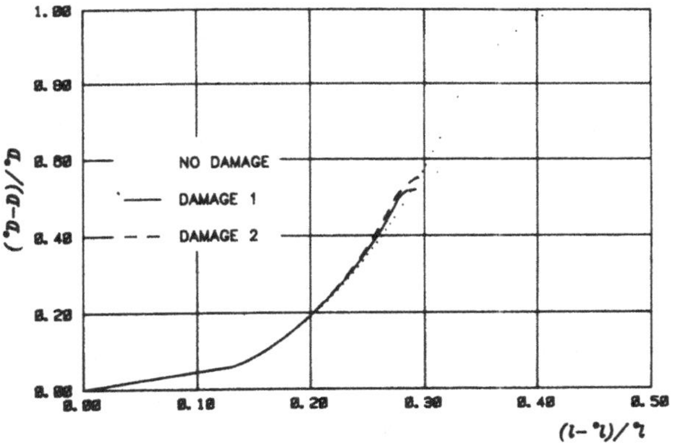

Fig. 41 Reduction of diameter at the neck.

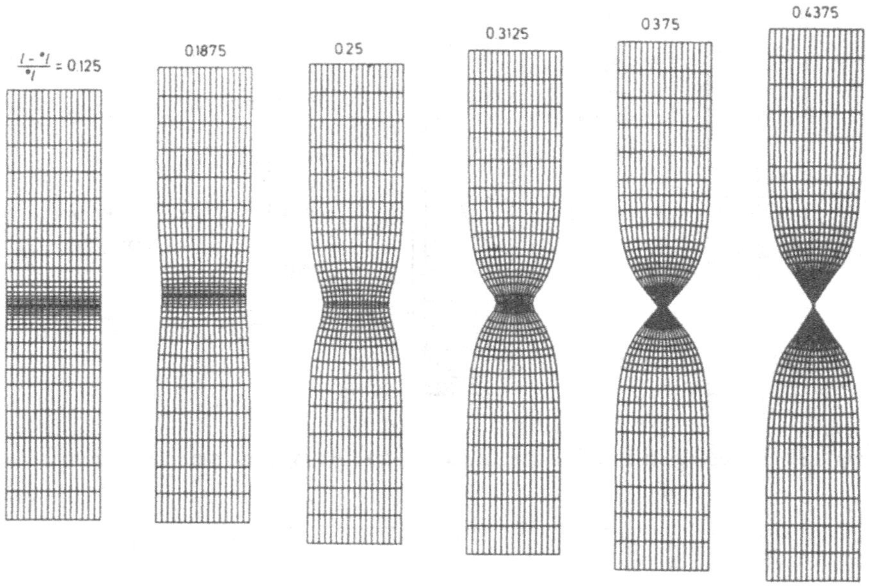

Fig. 42 Stages of deformation. Necking, no damage.

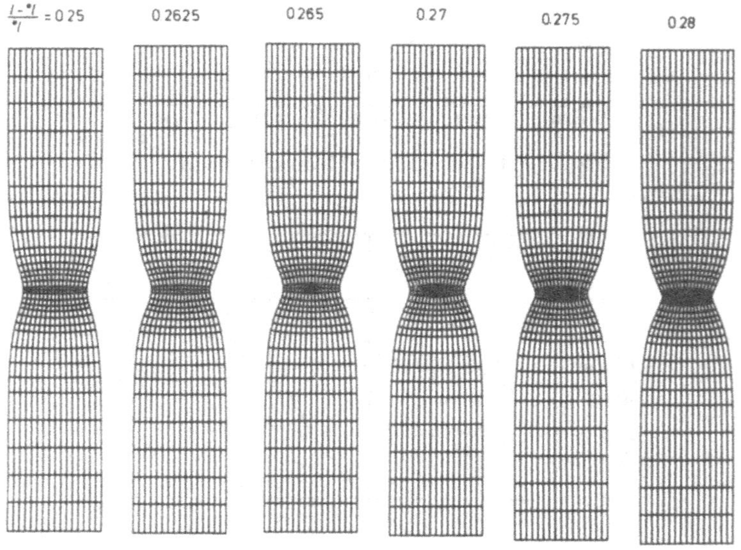

Fig. 43 Stages of deformation. Necking, damage and failure.

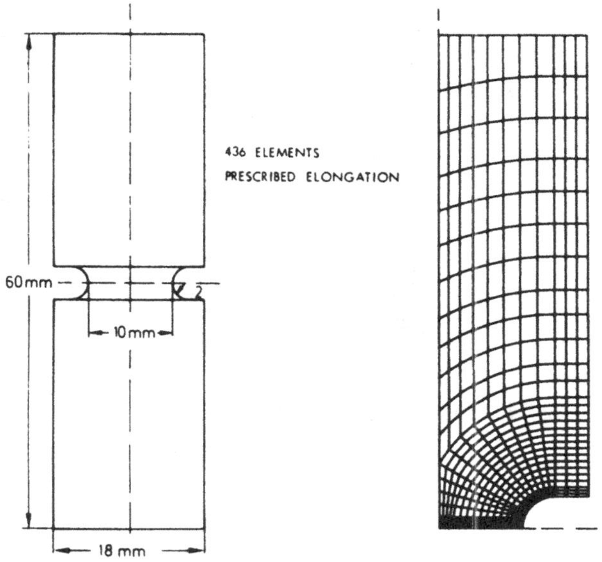

Fig. 44 Axisymmetric notched specimen.

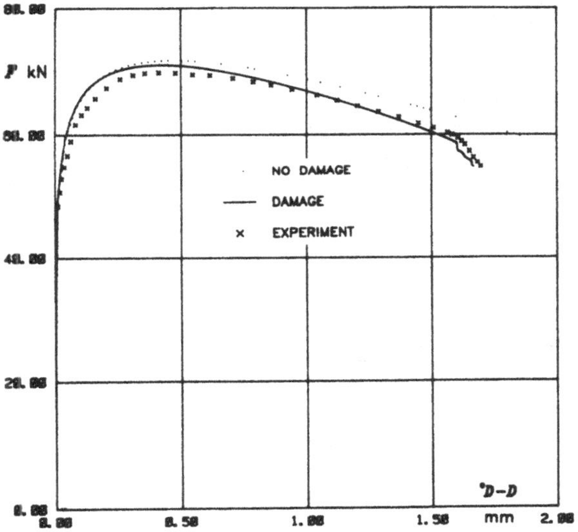

Fig. 45 Load versus contraction at the notch.

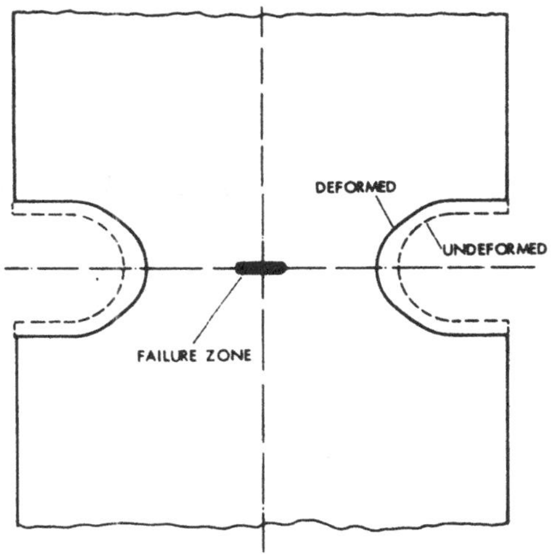

Fig. 46 Deformation of notch at incipient failure.

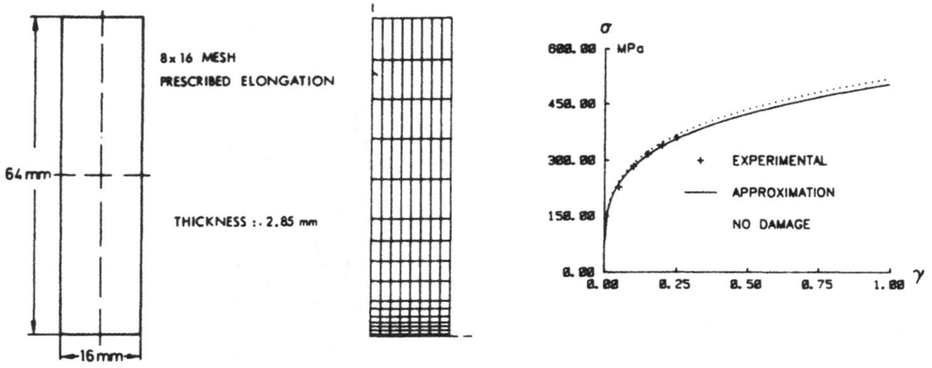

Fig. 47 Flat sheet under tension.

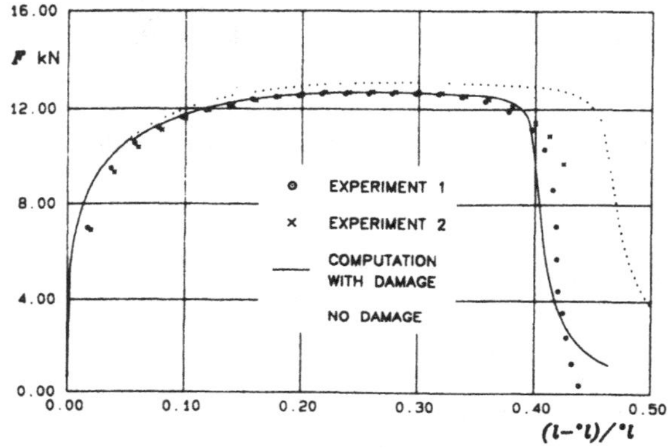

Fig. 48 Load – elongation diagram.

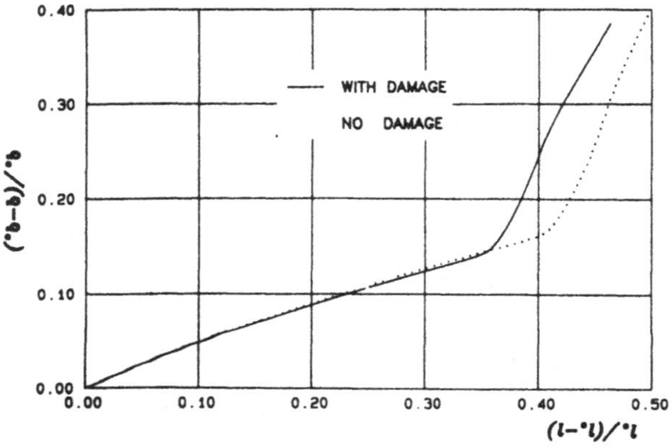

Fig. 49 Reduction of width at the neck.

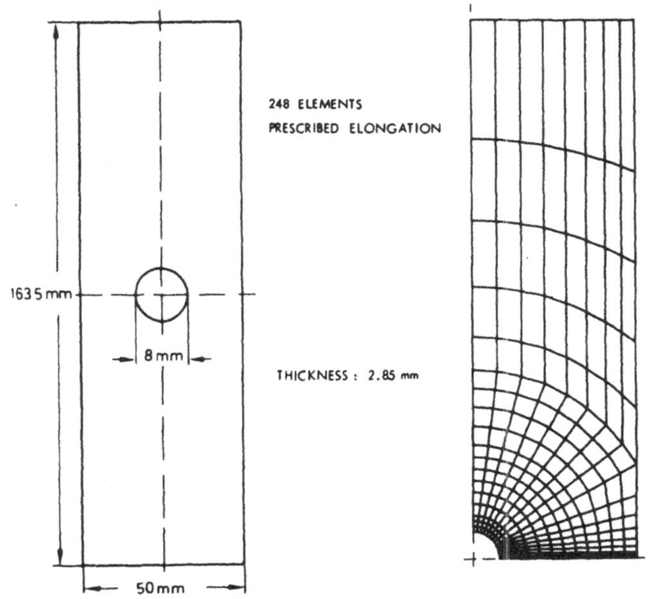

Fig. 50 Flat sheet with circular hole.

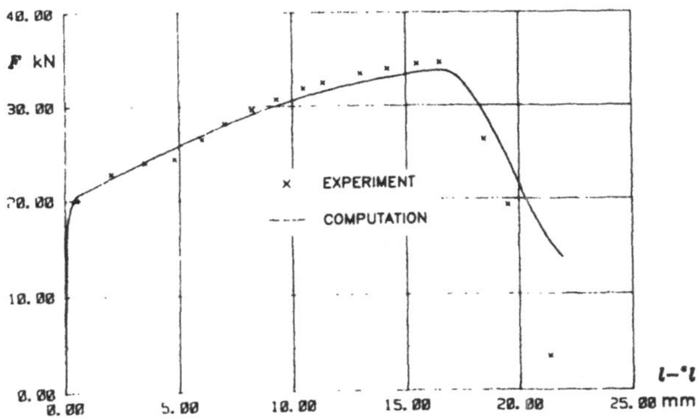

Fig. 51 Load – elongation diagram.

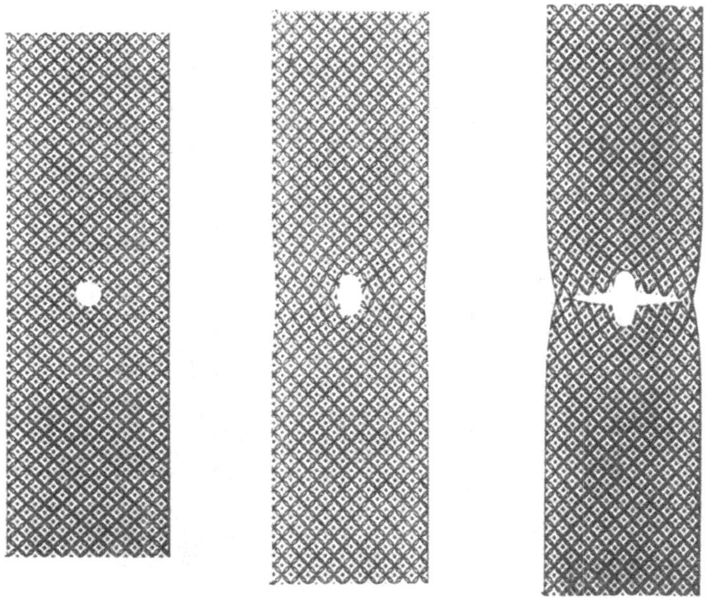

Fig. 52 Stages of deformation. Experiment.

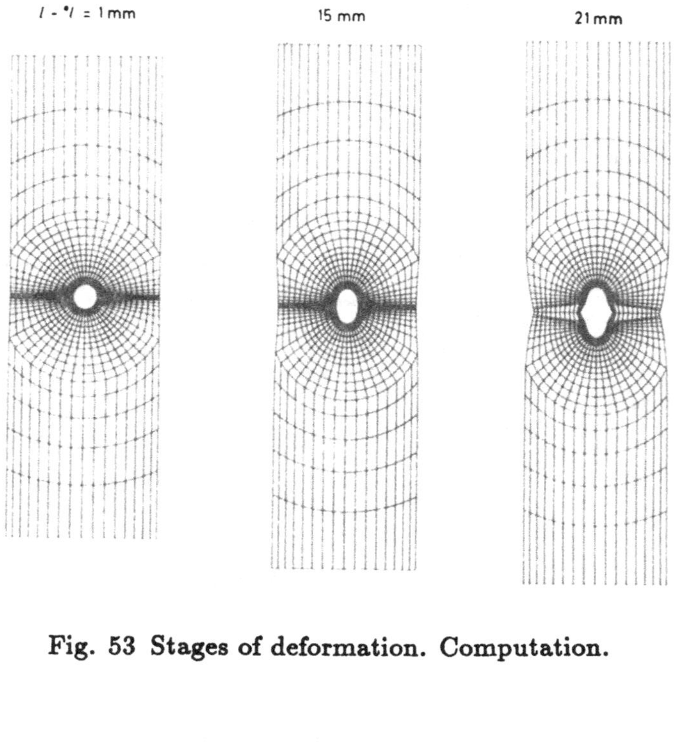

Fig. 53 Stages of deformation. Computation.

CHAPTER 2

COMPUTATIONAL TREATMENT OF TRANSIENT PROBLEMS IN NONLINEAR STRUCTURAL MECHANICS

J. P. Halleux
Joint Research Centre, Ispra, Italy

1. INTRODUCTION

Nonlinear transient problems are often encountered in structural dynamics. While vibration induced problems are usually linear or at most weakly nonlinear and can, therefore, be suitably solved in the frequency domain by applicatition of mode superposition techniques, many mainly accidental loading cases lead to important nonlinear phenomena and may involve wave propagation aspects. Both internal and external phenomena are to be considered: missile impact on industrial plants, shocks induced by fluids or gases during explosions or due to component failure in power plants for instance, crashes of cars or shipping casks, etc. Slower events such as seismic effects usually reduce the actual problem to a modal analysis although such statements are to be taken with some care. In any case the present lecture notes do not consider inertial problems but rather concentrate on phenomena with relatively short time scales (say of the order of milliseconds for the structures of common interest) which typically exhibit, at least in their initial phase, strong nonlinear effects. These problems require direct time integration techniques and the solution is thus incrementally constructed in the time domain via a step-by-step procedure.

Of course, the treatment of such structural problems involving at the same time transient loading, complex geometry, nonlinear history dependent material behaviour, large local rotations and large strains is often considered to be of a great complexity because of the numerous difficulties to be tackled at the same time. Though it is not wanted here to declare everything as being merely simple, it is, however, the intention to propose a general and at the end rather simple method which is believed to provide for excellent chances of success in tackling these problems. The method proposed is certainly not the only viable one but it lends itself to an easy numerical implementation on a computer, and it actually demonstrates to behave in a particularly robust and transparent manner. Only a minimum of user interaction is needed to run the type of codes then obtained. No obscure empirical parameters related to convergence or to ad hoc correcting terms, etc., are needed nor used. The procedure is remarkable also in that its derivation even in the geometrically nonlinear case does not make use of complex continuum mechanics aspects. The ingredients just match together in a quite natural way permitting physical insight and correlation with engineering common sense ... at least up to a certain point. Care should, for instance, still be observed with respect to the meaning of the constitutive relations used. There might be the feeling that currently existing methods are not often so limpid. Furthermore, it should be said that attempts to extend methods conceptually covering only a part of the difficulties enumerated in the beginning can become a hopeless torture in the sense that even if they come to work, they could well have lost their former cost effectiveness. It should, of course, also be remembered here that cost effectiveness is to be established not on the basis of the last or final run but rather on the whole series of runs and human effort needed to obtain an acceptable solution. On the other hand, pure theoretical considerations can be attractive from an intellectual point of view but they are vain if they do not cope with all the real ingredients of the problem. The challenging aspect appears rather to be the search for a possibly simple solution scheme to this class of problems keeping in mind that the goal is indeed to obtain a true predictive capability.

In order to arrive to think of a scheme of those considered here, it may be useful to remove first some of the usual clichés in the matter of time integration techniques. A largely diffused idea in this connection is that implicit methods are necessarily better than explicit methods, except maybe as far as simplicity is concerned. This sort of statement is indeed often taken for granted in many textbooks and most literature. In spite of that it is intended to express serious doubts in respect and suggest at least some caution in analysing these matters. In particular it will be seen that an explicit integration method forms a sound and convenient framework (i.e. not only with simplifying ideas in mind) for the solution of a large number of problems in the area of transient nonlinear mechanics.

Among the great advantages of an explicit time integration scheme is the possibility of testing equilibrium directly on the current configuration considering the true (i.e. Cauchy) stresses acting on the structure. This is possible by a mere application of the virtual work principle in its original form without any necessity for iterating or whatsoever.

Due to the presence of history dependent (elasto-plastic) material behaviour it is preferable to stick to the material points and a lagrangian mesh description is used. To put the equations into a time discretized form requires, however, attention and in particular for geometrically nonlinear problems this concern will be analysed in some depth. The integration of constitutive equations (stress update) is addressed in the frame of elasto-plastic modelling and also in respect to the presence of large rotations and large strains.

Before closing this introductory section, reference should be made to S.W. Key who provided dominant contributions for the set-up of the computational methods presented herein to deal with transient nonlinear structural problems. See, for instance, [1,2,3,4] of which much is also drawn here.

2. LAGRANGIAN-CAUCHY-STRESS FORMULATION

2.1 Kinematic preliminaries

The deformation gradient $\underline{F}$ is a matrix defined as

$$\underline{F} = \frac{\partial \underline{x}}{\partial \underline{X}} \tag{1}$$

where:
$\underline{X}$ is the position vector of a particle in the undeformed configuration;
$\underline{x} = \underline{x}(\underline{X},t)$ is the current position of the particle which stayed in $\underline{X}$ in the undeformed configuration (at time $t = t_o$).

A material point is thus carried from a spatial position $\underline{X}$ in a reference state at time t_o to a spatial position $\underline{x}$ at time t. $\underline{X}$ are the material coordinates and $\underline{x}$ the spatial coordinates. The sticking to material coordinates represents a lagrangian description of motion. Time t may be real time (dynamic problems) or a pseudo-time parameter characterizing the evolution of a problem (e.g. in relation to a proportional loading pattern). Coordinates are supposed to be referred to a fixed system of cartesian axes $(\underline{e}_1,\underline{e}_2,\underline{e}_3)$. For instance, the coordinates of vector $\underline{x}$ read x_1,x_2,x_3 and are generically denoted by x_j, $j = 1,2,3$.
A unique displacement vector, representing relative motion, can then be defined for each particle:

$$\underline{u}(\underline{X},t) = \underline{x}(\underline{X},t) - \underline{X} \tag{2}$$

The polar decomposition theorem permits to write:

$$\underline{F} = \underline{R}\,\underline{U} = \underline{V}\,\underline{R} \tag{3}$$

where:
$\underline{R}$ is a proper orthogonal rotation matrix;
$\underline{U}$ and $\underline{V}$ are the right and left stretch tensors, respectively.

A rate of rotation matrix $\underline{\Omega}$ is obtained by

$$\underline{\Omega} = \dot{\underline{R}}\,\underline{R}^T \tag{4}$$

where the dot denotes time derivative.

The velocity gradient $\underline{L}$ is a matrix defined as

$$\underline{L} = \frac{\partial \dot{\underline{x}}}{\partial \underline{x}} \tag{5}$$

The skew-symmetric part of $\underline{L}$ represents a rate of rotation matrix ($\underline{W}$) usually called spin, while the symmetric part is the rate of deformation ($\underline{D}$) usually called stretching tensor.

2.2 Equations of motion

If $\underline{x}$ is the spatial position where the stress is considered, it is easy to show that the equilibrium (maybe in a dynamic case, conservation of momentum would be a more adequate expression) equations in the current configuration read:

$$\rho \ddot{x}_i = \sigma_{ij,j} + f_i \qquad \text{in } V \tag{6}$$

$$\sigma_{ij} = \sigma_{ji} \tag{7}$$

where:

σ_{ij} are true stress components (Cauchy stress tensor);
f_i are the components of the body force density;
ρ is the mass density;
V is the volume occupied by the current configuration;
$\cdot$ (the superposed dot) denotes time derivative;
$,j$ designates derivative with respect to x_j.

Note that stress and deformation (rate) tensors will be represented in the following by simple Greek letters as usual, $\underline{\sigma}$ (sometimes $\underline{\alpha}$, $\underline{\eta}$ or $\underline{\xi}$) and $\underline{\varepsilon}$.

Furthermore, the boundary (S) natural conditions are expressed as

$$\sigma_{ij} n_j = t_i \qquad \text{on } S_1 \tag{8}$$

where:

n_j are the components of the unit normal to the surface at the boundary;
t_i are the components of the surface traction,

and essential boundary conditions as

$$x_i = \bar{x}_i \qquad \text{on } S_2 \tag{9}$$

where $\bar{x}_i$ are the components of the imposed motion.

2.3 Principle of virtual work and discretized equilibrium equations

A variational form of the equilibrium equations is obtained in terms of the principle of virtual work [1]:

$$\int_V \rho \ddot{x}_i \delta x_i dv = - \int_V \sigma_{ij} \delta x_{i,j} dv + \int_V \rho f_i \delta x_i dv + \int_{S_1} t_i \delta x_i ds \qquad (10)$$

The above is to hold for all variations δx_i satisfying the essential boundary conditions on S_2.

Via a finite element semidiscretization (for a detailed description see [5]), the following set of discrete differential equations in time can be obtained:

$$\underline{M}\ddot{\underline{u}} = \underline{f}^{ext} - \sum_e \int_{V^e} (\underline{B})^T \underline{\sigma} dv \qquad (11)$$

$\underline{\sigma}$ is now a matrix representing the approximate Cauchy stresses acting in the structure;
$\underline{M}$ is a mass matrix;
$\ddot{\underline{u}}$ is the nodal acceleration vector;
$\underline{f}^{ext}$ is a vector of externally applied loads;
$\underline{B}$ is the matrix of shape functions derivatives;
V^e represents the element (e) volume in the current configuration.

The set of equations (11) will result to be decoupled because the matrix $\underline{M}$ will be rendered diagonal by a suitable lumping process and the accelerations $\ddot{\underline{u}}$ will be directly obtained. In Section 3 is shown how to estimate $\underline{u}$ once $\ddot{\underline{u}}$ is known and in Sections 5 and 6, for the geometrically nonlinear case, how the stresses are obtained.

3. EXPLICIT TIME INTEGRATION ALGORITHM

3.1 Definition of the central difference scheme

It is assumed that all discretized quantities (displacement $\underline{u}$, velocities $\dot{\underline{u}}$, accelerations $\ddot{\underline{u}}$, stresses $\underline{\sigma}$ and related variables) are known at time t^n and denoted simply by the presence of the upper indice n. In order to obtain the same quantities at time $t^{n+1} = t^n + \Delta t$, where Δt is the time increment, it is first useful to introduce a modified velocity:

$$\underline{v}^{n+\frac{1}{2}} = \dot{\underline{u}}^n + \frac{\Delta t}{2} \ddot{\underline{u}}^n \qquad (12)$$

which is denoted $\underline{v}$ in order to stress the difference with $\dot{\underline{u}}$. The suffix $n+\frac{1}{2}$ indicates that $\underline{v}$ is a half-step approximation and is the constant velocity with which one moves from configuration n to configuration n+1 in the discretization process. The new displacements are thus given by:

$$\underline{u}^{n+1} = \underline{u}^n + \Delta t \, \underline{v}^{n+\frac{1}{2}} \qquad (13)$$

On this new configuration the stresses $\underline{\sigma}^{n+1}$ can now be evaluated by application of the constitutive relations (see Section 5 for the case of elasto-plastic material behaviour). These, in turn, permit via the discretized equilibrium Eqs.(11) to compute the new field of accelerations $\ddot{\underline{u}}^{n+1}$. Finally, new velocities are obtained as:

$$\dot{u}^{n+1} = \dot{u}^{n} + \frac{\Delta t}{2}\left(\ddot{\underline{u}}^{n} + \ddot{\underline{u}}^{n+1}\right) \tag{14}$$

The time integration scheme defined in the preceding equations is explicit in that all the quantities in the right-hand side terms are known when the equations are applied. No system solver is needed.

Often the central difference integrator is presented in the following form which, though obviously correct, tends to create some confusion regarding its explicit character:

$$\dot{\underline{u}}^{n+1} = \dot{\underline{u}}^{n} + \frac{\Delta t}{2}\left(\ddot{\underline{u}}^{n} + \ddot{\underline{u}}^{n+1}\right) \tag{15}$$

$$\underline{u}^{n+1} = \underline{u}^{n} + \Delta t\left(\dot{\underline{u}}^{n} + \frac{\Delta t}{2}\,\ddot{\underline{u}}^{n}\right) \tag{16}$$

It is indeed of utmost importance for the understanding of the methods proposed herein to note that the new configuration induced by the displacements $\underline{u}^{n+1}$ is obtained first, while the velocities $\dot{\underline{u}}^{n+1}$ which correspond to this configuration result from the application of the virtual work principle to this new configuration and are thus computed only at the end of the time stepping procedure.

The name of this scheme results from the following expression which is easily obtained in case of constant time step by introducing the displacement at time t^{n-1} and eliminating the velocities

$$\ddot{u}^{n} = \frac{1}{\Delta t^{2}}\left(u^{n-1} - 2u^{n} + u^{n+1}\right) \tag{17}$$

Finally, it may be recalled that no damping is introduced by the central difference operator, as can be seen by examining the form of the β-Newmark family of integrators [2] (i.e. $\gamma = \frac{1}{2}$).

3.2 Stability of the central difference scheme

Searching for a solution of (17), where the acceleration is substituted according to the basic equilibrium equation (11), neglecting external forces, under the form

$$\underset{\sim}{\delta}^{n} = \underset{\sim}{\delta}\, e^{\gamma t} = \underset{\sim}{\delta}\, e^{\gamma n \Delta t} \tag{18}$$

where $\underset{\sim}{\delta}$ is an arbitrary displacement shape and γ a parameter, an eigenvalue problem is obtained:

$$\left(e^{\gamma \Delta t} - 2 + e^{-\gamma \Delta t} - \Delta t^{2}[M]^{-1}[K]\right)\underset{\sim}{\delta} = 0 \tag{19}$$

To each eigenvalue ω_i^2 of $[M]^{-1}[K]$ corresponds an eigenvector δ_i and a particular value for γ. $[K]$ has the standard finite element meaning of a stiffness matrix [5]. For stability it is advisable to ensure

$$|e^{\gamma\Delta t}| \leq 1. \tag{20}$$

This leads to the classical stability limit:

$$\Delta t \leq \frac{2}{\omega_{max}} \tag{21}$$

where ω_{max} is the frequency which corresponds to the highest eigenvalue which depends on the spatial equations and discretization.

As recalled by Belytschko in [6], the maximum eigenvalue of a finite element system is, of course, bounded with sufficient accuracy by the maximum eigenvalue found in the single elements (e) and, therefore:

$$\omega_{max} \leq \max_{\text{for all } e} \omega_{max}^{e} \tag{22}$$

If a formula is known for estimating the stability limit of any finite element on the basis of its own characteristics and of the material model used, a global time step limitation is easily found or even better a map of local time step can be constructed in order to operate in a time partitioned manner (see section 7.3).

It is maybe useful also to settle an idea on the typical size of an admissible time step. For the linear wave equation ($u_{tt} = c^2 u_{xx}$) with a piecewise linear displacement finite element model of uniform element lengths Δx with lumped mass, the maximum frequency is given by:

$$\omega_{max} = 2\,\frac{c}{\Delta x} \tag{23}$$

which leads to

$$\Delta t \leq \frac{\Delta x}{c} \tag{24}$$

a result indicating that the critical time step corresponds to the time of travel of the longitudinal wave over the length of one element. This result also clearly shows that a balance is to exist between the discretization in time and space in order to obtain optimum responses and that this balance corresponds to a physical meaning which is easily caught by the engineer.

It should be noted, however, that though the order of magnitude of the critical time step is usually of that order for accurate schemes, it is not necessarily identical to expression (24). For instance, if mass lumping is not used and the consistent mass is retained, the maximum frequency is given by

$$\omega_{max} = \sqrt{6}\,\frac{c}{\Delta x} \tag{25}$$

and the stability condition becomes

$$\Delta t \leq \frac{2}{\sqrt{6}}\,\frac{c}{\Delta x} \tag{26}$$

4. CRITICAL ANALYSIS OF TIME AND SPACE DISCRETIZATIONS

4.1 Some comments related to explicit versus implicit time integration

As was recalled in the previous section, the central difference time integration scheme appears to be only conditionally stable. This means that a very large number of time steps may be necessary in practical cases. Indeed, for lagrangian meshes the stability condition can become computationally disastrous when the elements get strongly deformed (very short distance between nodes). This may then even require mesh rezoning for stability reasons and not only for accuracy reasons due to excessive skewing. Rezoning is a cumbersome and lousy operation for solids behaving in a memory dependent manner (like elasto-plasticity). Therefore, conditional stability is often viewed as being an unacceptable restriction and favour is given to implicit integration schemes of which many are unconditionally stable. However, a few other considerations need to be made to assess the respective advantages of explicit and implicit time integration schemes. For instance, it is important to realize that a large number of nonlinear problems require small time steps in order to achieve acceptable accuracy independently of stability considerations. In particular, wave propagation problems, i.e. those in which the behaviour of wave fronts is of engineering importance, require very small time steps at least locally in order to avoid frequency cutting and to catch correctly the position of the front and the often intricated nonlinear phenomena related to it. Typically, impact problems belong to this class though it must be kept in mind that each problem is a particular case and does not necessarily follow the general rule.

In practice what happens is that implicit integrators can do a fine job on high frequencies only if time steps of the order required by explicit integration are used. Obviously, the less high frequencies to be captured accurately, the better the implicit operator will do. But when these high frequencies become important, explicit integration is hard to beat for a number of reasons:

- as explicit methods are to be used with diagonal mass representation, as will be seen in the next paragraphs, in order to exhibit optimal spectral properties, no system solver is needed;
- when the mesh grows (large problems), the computational effort increases only proportionally to the number of degrees of freedom (no band width);
- nonlinearities are by far easier to treat: no convergence criteria for plasticity or large deformation, straightforward handling of fluid-structure coupling, of contact-impact problems, of friction boundary conditions (heterogeneous material like reinforced concrete), etc.;

- elegant algorithmic structure because of a simple logic;
- computationally attractive because of the few operations required per time step and the reduced core storage requirements;
- computationally effective because of the robustness of the explicit time integration procedure: completion of a computation does not require any tuning and reliability of the results can be assessed by a simple energy check.

4.2 Spectral properties of time integrators and effect of mass representation

In order to understand the behaviour of a time integration scheme it is interesting to analyse the ratios between theoretical frequencies and the correspondingly obtained numerical approximations. The β-Newmark family is considered in [2] and leads to interesting results which are plotted in Fig. 1. $\beta = 1/4$ corresponds to the trapezoïdal rule (average acceleration method) which is a commonly used implicit integrator, while $\beta = 1/6$ for instance is the linear acceleration method also of an implicit character. The reason that $\beta = 1/4$ is often advocated is that it leads to unconditionally stable behaviour but it is also easily seen on Fig. 1 that the frequency response is rapidly degraded when the time step increases. Indeed, the simple explicit central difference scheme offers good behaviour with respect to the unconditionally stable $\beta = 1/4$ integrator.

Particularly important to note here also is that frequencies are overestimated with the use of the central difference operator, while they are underestimated with the trapezoïdal rule. This leads, as will be shortly seen, to very important considerations.

Consider now the effect on frequency response due to the spatial finite element modelling, more precisely the mass representation, which can be "consistent" (standard finite element procedure) or "lumped" (rendered diagonal for instance by summing the terms on the diagonal), or something in between, as will be seen later. Again following [2], results are plotted in Fig. 2. Again one representation increases the frequencies (consistent mass representation), while the other one depresses frequencies (diagonal mass). The scope of this paragraph now becomes clear: it is obvious that some combinations are better than others and indeed similar as far as computed results are concerned. These are explicit time integration with diagonal mass and implicit (at least for trapezoïdal rule) with consistent mass which is indeed the most standard finite element approach. Diagonal mass with $\beta = 1/4$ implicit integration would deliver poor frequency response as has often been observed in the literature, unfortunately to the disadvantage of mass lumping techniques. The nicest with this situation is the kindness of nature for the particular combination of explicit time integration and lumped mass:

- the system of equilibrium equations (11) is then immediately solved with no need for matrix inversion;
- the stability limit which results from the use of lumped mass is necessarily larger than the one which would result with consistent mass.

Recalling the argumentation once again the following summarizes the general guidelines followed in setting up a solution procedure. In the discretization process of the original partial differential equations, approximations are induced both by the finite element modelling in space and by the use of a time integrator to advance the solution in time. The scope being that actual computed results should show up as a good approximation to the exact solution of the original problem, it appears to be useful to introduce a concept of balance between time and space discretization. It is not necessary that for both the space modelling and the time integration to be on an individual basis excellent but it is sufficient that their combined effects deliver good results. Compensation may be the adapted term here. Errors due to the respective approximations in space and time do not necessarily show up in the global approximation; they may indeed be mitigated. In this respect, for instance, care is also to be taken when a high order integration scheme is to be applied in connection with a poor spatial representation: the better integration of high frequencies could then lead to a clear disadvantage because this part of the spectrum is anyhow largely biased by the insufficient representation in space.

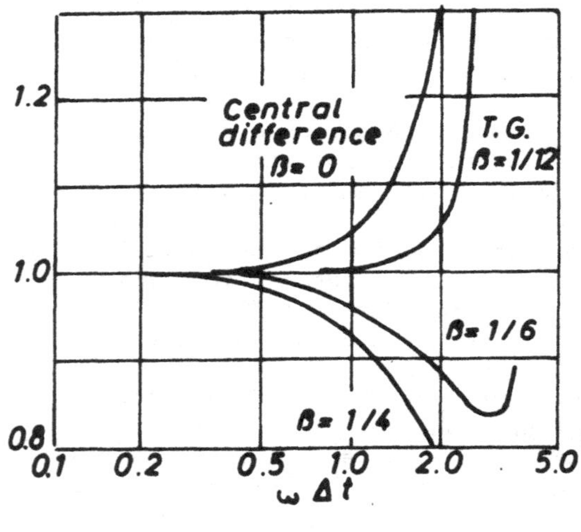

Fig. 1. Frequency response for $\gamma = \frac{1}{2}$ for the Newmark β time integrator.

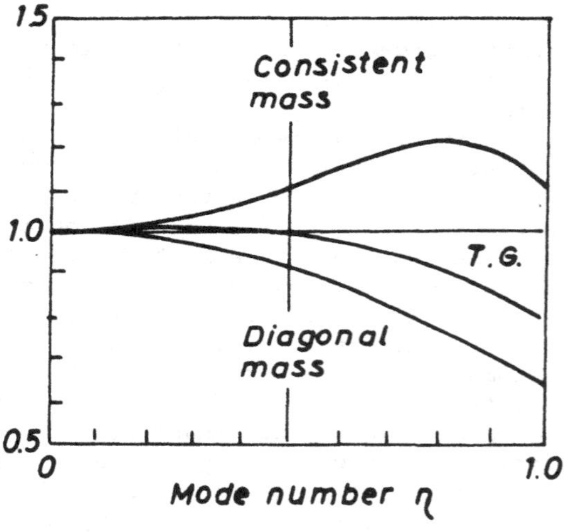

Fig. 2. Frequency response for various mass representations.

4.3 A high order scheme for wave propagation analysis

In order to demonstrate that implicit integration is not necessarily synonymous with unconditional stability, and even more, that the stability condition related to a lumped mass, explicit time integration has a profound meaning, a 4th-order Taylor-Galerkin scheme is now derived for the one-dimensional wave equation already mentioned in Section 3.2:

$$u_{tt} = c^2 u_{xx} \tag{27}$$

A 4th-order accurate time stepping method for Eq.(27) can be obtained by a combination of forward and backward Taylor expansions in time which yields

$$\frac{u^{n+1}-2u^{n}+u^{n-1}}{\Delta t^2} = u^n_{tt} + \frac{\Delta t^2}{12} u^n_{tttt} + O(\Delta t^4) \tag{28}$$

or, using Eq.(27)

$$\frac{u^{n+1}-2u^{n}+u^{n-1}}{\Delta t^2} = c^2 u^n_{xx} + \frac{c^2 t^2}{12} \partial^2_x\left(u^n_{tt}\right) + O(\Delta t^4) \tag{29}$$

Now, in view of (28):

$$\partial^2_x\left(u^n_{tt}\right) = \partial^2_x\left(\frac{u^{n+1}-2u^{n}+u^{n-1}}{\Delta t^2}\right) - \frac{\Delta t^2}{12} \partial^2_x\left(u^n_{tttt}\right) + O(\Delta t^4) \tag{30}$$

Therefore, the 4th-order time accuracy is maintained if Eq.(29) is replaced by

$$\left(1 - \frac{c^2\Delta t^2}{12} \partial^2_x\right)\left(\frac{u^{n+1}-2u^{n}+u^{n-1}}{\Delta t^2}\right) = c^2 u^n_{xx} \tag{31}$$

It is readily seen that scheme (31) is nothing but a β-Newmark scheme [2] with $\beta = 1/12$ and $\gamma = 1/2$ (no damping). Fig. 1, extracted from [2], shows the frequency response for $\gamma = 1/2$ of the Newmark β time integrators and clearly confirms that optimal properties are obtained with $\beta = 1/12$ for problems involving a wide spectrum of frequencies.

To produce a fully discrete system, the Galerkin formulation is applied to the semi-discrete equation (31). Assuming a uniform mesh of linear elements of size h, this gives the following discrete equation at an internal node j:

$$\left[1 + \left(r - \frac{c^2\Delta t^2}{12h^2}\right)D\right]\ddot{u}_j - \frac{c^2}{h^2} D\, u^n_j = 0 \tag{32}$$

where $r = 1/6$, $\ddot{u}_j = \left(u^{n+1}_j - 2u^n_j + u^{n-1}_j\right)/\Delta t^2$, and $D\,u_j = u_{j+1} - 2u_j + u_{j-1}$.

An analysis of Eq.(32) reveals that

$$\left[1 - \frac{c^2\Delta t^2}{12h^2} D\right]\ddot{u}_j = \frac{\partial^2 u}{\partial t^2} + O(\Delta t^4) \tag{33}$$

which confirms the 4th-order temporal accuracy, while the remaining terms can be rewritten as

$$r D\bar{u}_j - \frac{c^2}{h^2} Du_j^n = -c^2 \frac{\partial^2 u}{\partial x^2} + c^2 h^2 \left(r - \frac{1}{12}\right) \frac{\partial^4 u}{\partial x^4} + O(\Delta t^2 h^2, h^4) \qquad (34)$$

It follows that the standard Galerkin formulation ($r = 1/6$, i.e. consistent mass) produces a scheme which is only 2nd-order accurate in the mesh size. Furthermore, the stability condition of scheme (32) with $r = 1/6$ is found to be $\alpha = c\Delta t/h \leq 1/\sqrt{3}$, a significant reduction with respect to the standard central difference scheme based on a diagonal mass representation which is stable for $\alpha \leq 1$. However, Eq. (34) indicates that a 4th-order spatial accuracy is obtained when $r = 1/12$. Moreover, with this choice the stability condition of the explicit central difference operator, $\alpha \leq 1$, is recovered. This last finding is important in that it shows that an accurate implicit operator is bound by a condition quite similar to that of the simple operator proposed here.

In practice, scheme (32) with $r = 1/12$ can be obtained by using a modified mass matrix consisting of the average between the consistent and the diagonal mass matrices.

Also for the spatial part of the Taylor-Galerkin scheme (30) with $r = 1/12$, an excellent frequency response is obtained, as can be observed from Fig. 2 which gives the ratio of discrete frequency to exact frequency versus the mode number for various mass representations. For $r = 1/12$ a virtually flat response is obtained, except for the highest mode numbers where the predicted frequencies are too low. Notice, however, that the time discretization (Fig. 1 with $\beta = 1/12$) predicts frequencies which are too high and, therefore, the proposed Taylor-Galerkin scheme achieves an optimum matching of spatial and temporal discretizations.

This Taylor-Galerkin scheme is interesting for pedagogical reasons but appears to be too complicated for current practical purposes where nonlinear material behaviour is systematically encountered and where generalization to 2-D and 3-D space is needed.

5. ELASTO-PLASTIC MATERIAL BEHAVIOUR

5.1 Classical approach to elasto-plastic material modelling

The equations mentioned in the previous sections have reduced the original problem to the development of a suitable algorithm for stress computation: once a new configuration has been obtained via the time integrator, it is necessary to evaluate the new stresses in order to be able to apply again the equilibrium equations (11). In the presence of elasto-plastic material behaviour this is easily achieved by taking advantage of the natural rate form description of such materials. Indeed, their constitutive equation, where the superscript p stays for plastic part, reads:

$$\dot{\sigma}_{ij} = C_{ijmn}\left(\dot{\varepsilon}_{mn} - \dot{\varepsilon}^p_{mn}\right) \qquad (35)$$

where C_{ijmn} is the elastic matrix (generalized Hooke's law) depending only on Young's modulus (E) and Poisson's ratio (ν)

$$C_{ijmn} = \mu(\delta_{im}\delta_{jn} + \delta_{in}\delta_{jm}) + \lambda\delta_{ij}\delta_{mn} \tag{36}$$

with the Lamé constants λ and μ (also called shear modulus) defined by

$$\lambda = \frac{E\nu}{(1+\nu)(1-2\nu)}, \qquad \mu = \frac{E}{2(1+\nu)} \tag{37}$$

As long as no permanent straining occurs, the material is said to behave elastically $\left(\dot{\varepsilon}^p_{ij} = 0\right)$. When the stress state is such that an infinitesimal change in stress can produce some permanent strains, the material has reached its elastic limit. The collection of these stress states is supposed to define a convex surface in the space of stresses which is called yield surface or plasticity surface. The admissible states of stress are those interior to, or on this surface. A yielding criterion is a law defining explicitly a yield surface. The Von Mises yield criterion reads

$$f(\sigma_{ij}) = \bar{\sigma} - \sigma_y = 0 \tag{38}$$

f being the yield function, and

$$\bar{\sigma} = \sqrt{\frac{3}{2} S_{ij}S_{ij}} \tag{39}$$

is called equivalent, effective or Von Mises stress, S_{ij} being the deviatoric stress state. In Eq.(38) σ_y is the elastic or yield limit in a tension test. The Von Mises surface can be represented in principal stress space as a cylinder oriented symmetrically with respect to the three axes (isotropy) and defined by its radius:

$$R = \sqrt{\frac{2}{3}}\, \sigma_y \tag{40}$$

An associative flow rule relating the incompressible $\left(\dot{\varepsilon}^p_{ii} = 0\right)$ plastic strain rate to the current stress state and strain rate, is further postulated:

$$\dot{\varepsilon}^p_{ij} = \dot{\Lambda}\frac{\partial f}{\partial \sigma_{ij}} \tag{41}$$

A consistency condition [7] is invoked to define $\dot{\Lambda}$ in such a way that, during plastic deformation, the stress state remains located on the yield surface.

For convenience, a trial rate of stress $\left(\dot{\sigma}^{tr}_{ij}\right)$ is introduced:

$$\dot{\sigma}^{tr}_{ij} = C_{ijmn}\dot{\varepsilon}_{mn} \tag{42}$$

For stress states lying inside the yield surface, the deformation process is, as already said, purely elastic $\left(\dot{\varepsilon}^p_{ij} = 0\right)$. For stress states lying on the surface, two cases are to be distinguished: if the trial rate of stress is pointing towards the interior of the yield surface (with reference to the plane tangent to the yield surface at the current state of stress), the process is again purely elastic $\left(\dot{\varepsilon}^p_{ij} = 0\right)$, otherwise the process involves permanent straining and is said to be plastic, $\dot{\varepsilon}^p_{ij}$ being defined by Eq.(41).

A hardening rule explicitly defines the evolution of the yielding surface during a plastic deformation process. The choice of a specific hardening rule depends on mathematical consistency with the yield function, on ease of numerical implementation, as well as on experimental evidence. Most commonly, a hardening rule conserving isotropy is considered: the yielding surface grows uniformly and keeps its original shape

$$F\left(\sigma_{ij}, \varepsilon^p_{ij}\right) = \bar{\sigma} - \sigma_y\left(\varepsilon^p_{ij}\right) = 0 \tag{43}$$

Linear isotropic hardening is represented by:

$$\dot{\sigma}_y = H\,\bar{\dot{\varepsilon}}^p_{ij} \tag{44}$$

where H is the plastic modulus, a material property, and $\bar{\dot{\varepsilon}}^p_{ij}$ is the equivalent plastic strain rate defined by:

$$\bar{\dot{\varepsilon}}^p_{ij} = \sqrt{\frac{2}{3}\,\dot{\varepsilon}^p_{ij}\dot{\varepsilon}^p_{ij}} \tag{45}$$

A given elasto-plastic stress state then always corresponds to the same yielding surface independently of the deformation history. In a uniaxial stress situation this means that, when reversed loading is applied (e.g. compression after a tension test), further plastic deformation takes place only when the compressive stress becomes equal to the last tension stress that produced plastic deformation.

From an experimental point of view the isotropic model is realistic only as long as no reversed loading is applied: in this case the Bauschinger effect (apparent reduction of the yield point if loading is reversed after some plastic deformation has occurred) has to be taken into account. This phenomenon destroys the initial isotropy of the material. For this effect to be taken into account, a "kinematic" hardening rule is designed: the yield surface moves unchanged in stress space

$$F = \sqrt{\frac{3}{2}\left(S_{ij}-\alpha_{ij}\right)\left(S_{ij}-\alpha_{ij}\right)} - \sigma_o = 0 \tag{46}$$

where σ_o is the initial yield stress and α_{ij} is called the back stress and represents the translation of the yield surface (initially equal to zero). Linear kinematic hardening is then described by:

$$\dot{\alpha}_{ij} = \frac{2}{3} H \dot{\varepsilon}^{p}_{ij} \tag{47}$$

A simple linear combination of isotropic and kinematic hardening then permits to simulate, to some extent even fairly realistically, the behaviour of materials submitted to arbitrary loading.

Using direct notation in which the dot between expressions denotes contracted (dot) product and calling β the parameter of proportionality between isotropic and kinematic hardening ($\beta = 0$ fully kinematic and $\beta = 1$ fully isotropic), a summary of the theory is presented in Table I. Following [7] an evolution problem can then be stated in the form: "given initial values of $\underline{\sigma}$, σ_y and $\underline{\alpha}$, and a time history of $\underline{\varepsilon}$, find the time histories of $\underline{\sigma}$, σ_y and $\underline{\alpha}$ such that the equations in Table I are satisfied". An interesting point to note here is that the present theoretical approach accommodates not only hardening but also softening behaviour [7].

5.2 The radial return algorithm

The basic idea lying in the radial return concept is first to estimate a trial stress by summing the old stress to an elastically induced stress increment based on the given strain increment and then, in the case where the new stress lies outside the yield surface, to bring the trial stress to the yield surface along a line pointing towards the centre of the yield surface in deviatoric space. This simplistic method has proven over the years not only to be the most attractive from a computational (implementation + cost) point of view, but also as far as accuracy is concerned.

An elegant generalization to the case of combined kinematic and isotropic hardening is obtained [3] by introducing an approximation to the vector normal to the yield surface. Again following the excellent summary proposed in [7]: the basic idea is to approximate the normal vector $\underline{N}$ by $(\underline{\xi}^{tr}_{n+1})'/|(\underline{\xi}^{tr}_{n+1})'|$, where $(\underline{\xi}^{tr}_{n+1})' = (\underline{\sigma}^{tr}_{n+1} - \underline{\alpha}_n)'$. The consistency condition is then applied in the large to the time-discrete counterparts of (35), (44) and (47) (or the first three equations in Table I):

$$\underline{\sigma}_{n+1} = \underline{\sigma}^{tr}_{n+1} - 2\mu \tilde{\Lambda} \underline{N} \tag{48}$$

$$R_{n+1} = R_n + \frac{2}{3} \beta H \tilde{\Lambda} \tag{49}$$

$$\underline{\alpha}_{n+1} = \underline{\alpha}_n + \frac{2}{3} (1-\beta) H \tilde{\Lambda} \underline{N} \tag{50}$$

This results in

$$\tilde{\Lambda} = \frac{1}{2\mu} \frac{1}{\left(1 + \frac{H}{3\mu}\right)} \left(\left| (\underline{\xi}^{tr}_{n+1})' \right| - R_n \right) \tag{51}$$

The algorithm is summarized in Table II. For a detailed description see [7].

6. LARGE ROTATION AND LARGE STRAIN PROBLEMS

6.1 Generalities

As was seen in the previous section, stress incrementation proceeds in a rather straightforward manner. When geometrically nonlinear effects are present, some extra care should be taken in handling the constitutive relations. For instance, when large rotations occur it is important to express the stress components correctly, in general by applying a suitable coordinate transformation or more simply maybe to stay in a local system rotated by the material motion. This latter procedure is useful mainly for thin members and requires a reprojection of the internal force vectors at the assembly stage. In the more general case of arbitrarily deforming bodies, it is, however, difficult, if not impossible, to distinguish in a physically sound manner (i.e. based on standard geometrical considerations) between what are mere rotation effects and what causes actual straining and thereby a modification in the stress state already reached by the material. To find a way out of this uncomfortable situation and possibly an easy one, is the scope of so-called "objective stress rate" approaches. In reality, this is nothing other than a casting of the constitutive equations into a rate form which guarantees invariance with respect to the observer (principle of objectivity) and under superposed coordinate transformations. The unfortunate aspect of these approaches is that a myriad of objective stress rates can be obtained and that the choice between them heavily relies on essentially crude interpretations of the kinematics involved. This then tends eventually to help in hiding the real difficulties involved in arbitrarily large straining phenomena. Nevertheless, when the deforming processes remain bounded in some way, it appears possible to set guidelines regarding the use of objective stress rates and the associated kinematic quantities.

6.2 Definition of a reference case

In order to evaluate the effects of the use of various stress rates, it is often proposed to consider the following elementary shear test:

$$u_1(t) = T(t)x_2; \quad u_2(t) = u_3(t) = 0 \tag{52}$$

where x_i and u_i are components of cartesian space coordinates and of displacements, respectively.

Based on finite elasticity considerations an analytical solution, where μ is the shear modulus (see Section 5) can be obtained [8]

$$\sigma_{11} = \frac{2T\mu}{\sqrt{T^2+4}} \ln\left(\frac{T+\sqrt{T^2+4}}{2}\right) \tag{53}$$

$$\sigma_{12} = \frac{4\mu}{\sqrt{T^2+4}} \ln\left(\frac{T+\sqrt{T^2+4}}{2}\right) \tag{54}$$

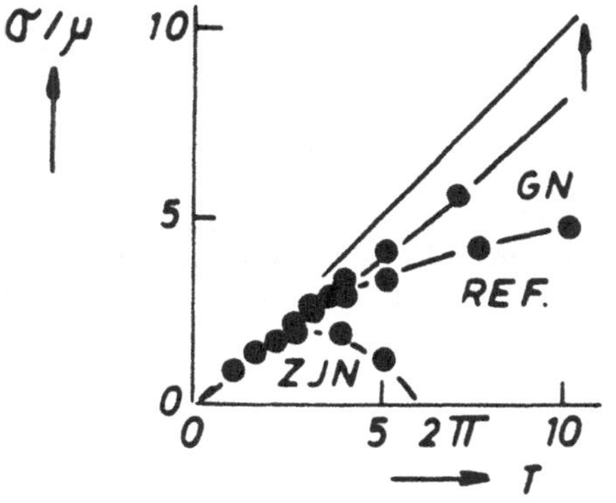

Fig. 3. Principal stress.

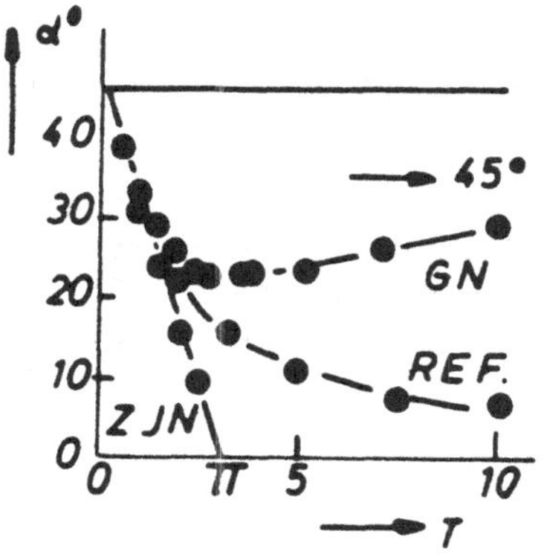

Fig. 4. Principal direction.

This solution is plotted as reference solution on Figs. 3 and 4.

The elasto-plastic material modelling defined in the previous section pertains, however, to a class of hypo-elastic rate constitutive equations and even in the elastic limit does not reduce to classical finite elasticity which is not based on rate forms. It seems, nevertheless, interesting in order to check the soundness of rate solutions, to compare them with the analytical solution given above.

6.3 The Jaumann rate of Cauchy stress

A hypo-elastic material [9] can be defined as behaving in the following way:

$$\overset{\nabla}{\underline{\sigma}} = \underline{C}(\underline{\sigma})\underline{D} \tag{55}$$

where the rate $\overset{\nabla}{\underline{\sigma}}$ called the Zaremba-Jaumann-Noll (ZJN) rate of Cauchy stress, unlike the Cauchy stress, is objective and given by:

$$\overset{\nabla}{\underline{\sigma}} = \dot{\underline{\sigma}} - \underline{W}\underline{\sigma} + \underline{\sigma}\underline{W} \tag{56}$$

The velocity gradient ($\underline{L}$):

$$\underline{L} = \underline{D} + \underline{W} \tag{57}$$

furnishes - as already stated in Section 2 - the rate of deformation in its symmetric part and the rate of rotation in its skew-symmetric part.

It may be noted that the ZJN stress rate ($\overset{\nabla}{\underline{\sigma}}$) is linear in the stretching ($\underline{D}$) and that the material response function ($\underline{C}$) depends on the stress ($\underline{\sigma}$) only. If $\underline{C}$ is further restricted to be stress-independent, a hypo-elastic material of grade zero (Jaumann type) is obtained. $\underline{C}$ can then be viewed as a generalized Hooke matrix depending on the elastic constants only. This is what offers an opportunity to include elasto-plastic material behaviour by merely replacing the elastic matrix as classically used in small strain analysis. As this depends on the stress state only, a form coherent with Eq.(55) is apparently obtained. However, theoretical considerations [9] certainly oblige to handle hypo-elastic materials of grade 0, or their generalization, with care.

As a matter of fact, a number of anomalies can be found when use is made of computational approaches based on this sort of material modelling. For instance, the shear test as defined by Eqs.(52) and with a constitutive relation such as (55), where $\underline{C}$ is the elastic matrix, delivers the following set of relations after application of (56):

$$\dot{\sigma}_{11} - \dot{T}\sigma_{12} = 0 \tag{58}$$

$$\dot{\sigma}_{22} + \dot{T}\sigma_{12} = 0 \tag{59}$$

$$\dot{\sigma}_{12} + \frac{1}{2}\dot{T}(\sigma_{11}-\sigma_{22}) = \dot{T}\mu \tag{60}$$

A periodic response is obtained

$$\sigma_{11} = -\sigma_{22} = \mu(1-\cos T) \tag{61}$$

$$\sigma_{12} = \mu \sin T \tag{62}$$

Principal value and angle are plotted in Figs. 3 and 4. This result is usually considered to be unsatisfactory from a physical point of view because of the oscillatory nature of the stresses. Among the remedies which have been proposed to this situation, the approach by Dienes [10] consisting of using a modified stress rate (called Green-Nagdhi) but still based on the stretching, has generated particular enthusiasm; this even to the point that many computer programs originally based on the ZJN rate switched to using this rate. The desirability of this abrupt change was critically examined in [8], where it was found that the use of this new rate is less convincing than numericists would like it to be. The Green-Nagdhi rate will be summarized in Section 6.4.

From a numerical point of view it is important to note that a straightforward use of Eq.(56) preserves objectivity in an infinitesimal sense, i.e. objectivity is achieved only in the limit of vanishingly small time steps. From geometrical considerations an elementary algorithm preserving objectivity for an arbitrarily large time step can be designed in 2-D. If the rotation angle during a time step is denoted by θ, a tensor $\underline{\tau}$ (the stress for instance) transforms like: $(\underline{T})^T\underline{\tau}\underline{T}$ with

$$\underline{T} = \begin{bmatrix} \cos\theta & \sin\theta \\ -\sin\theta & \cos\theta \end{bmatrix} \tag{63}$$

In the discrete process it is easily seen that the angle θ is related to the mid-point spin w_{12} by:

$$\Delta t\, w_{12} = 2 \tan \frac{\theta}{2} \tag{64}$$

This discrete procedure for preserving objectivity is equivalent in 2-D to a general procedure based on polar decomposition rather than on separation between the symmetric and anti-symmetric parts of the velocity gradient. It should be noted, however, that in practice direct use of Eq.(56) delivers satisfactory results for all applications where the rotation increments remain small. When a discretization process is applied, it is necessary to define how to compute the approximate strain rate (in fact, how strain incrementation is performed). It can be shown [3] that the best approximation for strain rate is obtained by making use of the stretching obtained at n+½. In particular, the accumulated strain in the absence of rotation then provides for an excellent approximation to the natural logarithmic strain which is of great advantage in the interpretation of large strain constitutive relations.

Going further into the discretization process and in order to preserve maximum accuracy in the application of the constitutive relations, a double transformation can be performed. First the stress is rotated by an angle θ/2 to the configuration n+½, then the constitutive relation is applied and finally the updated stress is transformed to configuration n+1. This procedure thus requires a double isoparametric transformation in a single step: one for configuration n+½ and one for configuration n+1.

The half angle transformation matrix is constructed according to Eq.(64) as:

$$\cos \frac{\theta}{2} = \frac{1}{\sqrt{1 + \frac{w_{12}^2 \Delta t^2}{4}}} \quad ; \quad \sin \frac{\theta}{2} = \frac{\Delta t\, w_{12}/2}{\sqrt{1 + \frac{w_{12}^2 \Delta t^2}{4}}} \tag{65}$$

Using half step approximations for the stretching and spin, and implementing the constitutive relations in the way described here, nearly exact solutions can be obtained [3].

6.4 The Green-Nagdhi rate of the Cauchy stress

The Green-Nagdhi stress rate, GN [10], is based on mixed considerations taken on the one hand from hypo-elasticity and on the other from finite elasticity. A hypo-elastic-like constitutive equation is used relating some objective stress rate ($\overset{\vee}{\underline{\sigma}}$) to the stretching ($\underline{D}$):

$$\overset{\vee}{\underline{\sigma}} = \underline{C}(\underline{\sigma})\underline{D} \tag{66}$$

As in hypo-elasticity the rate of deformation ($\underline{D}$) is thus taken as the measure of strain rate. The stress rate ($\overset{\vee}{\underline{\sigma}}$) is, however, not related to the rate of Cauchy stress via the spin ($\underline{W}$) as in (56), but via the rate of rotation ($\underline{\Omega}$) which is derived (4) from the proper orthogonal matrix ($\underline{R}$) obtained in the polar decomposition (3) of the deformation gradient ($\underline{F}$):

$$\overset{\nabla}{\underline{\sigma}} = \dot{\underline{\sigma}} - \underline{\Omega}\underline{\sigma} + \underline{\sigma}\underline{\Omega} \tag{67}$$

Despite the critics which have been raised [8] against this approach the set of equations (4, 66, 67) can be chosen to define some sort of generalized hypo-elastic [11] behaviour. To complete the analysis attempted here of the Green-Naghdi stress rate approach, it remains to check how this affects the solution of the test case which was chosen. Thus, turning again to the shear test, the following solution is obtained [10] in terms of the angle of material rotation (θ) defined by:

$$\theta = \tan^{-1} \frac{T}{2} \tag{68}$$

$$\sigma_{11} = -\sigma_{22} = 4\mu(\cos 2\theta \, \ell n \cos\theta + \theta \sin 2\theta - \sin^2\theta) \tag{69}$$

$$\sigma_{12} = 2\mu \cos 2\theta (2\theta - 2\tan 2\theta \, \ell n \cos\theta - \tan\theta) \tag{70}$$

Principal value and angle are again found in Figs. 3 and 4.

While at first sight the GN solution certainly looks better than the ZJN solution, it does not appear to present a very sound behaviour for large values of T (θ approaching $\pi/2$).

Despite the presence of logarithms the limiting behaviour at infinity is identical to a purely linear solution, i.e. principal directions at $\pm$ 45° and linear dependence on T. Consistent behaviour, however, should yield exactly the opposite, i.e. principal directions at 0 and 90° and, of course, true logarithmic dependence on T. As a matter of fact, rapid deviation from the analytical solution is observed as soon as T exceeds 1, i.e. when ZJN solutions start to exhibit their oscillating character. GN solutions are, therefore, not felt as being exceptionally remarkable and hardly can justify a large superiority over ZJN solutions.

While for the Jaumann stress rate one uses an incremental rotation of the stress at each step, ideally obtained by polar decomposition, the rotated Cauchy stress method uses the polar decomposition theorem applied to the deformation gradient. A numerical procedure is easily derived by making explicit use of the orthogonal rotation matrix obtained from the deformation gradient. The computational effort needed appears, however, to be larger than for the ZJN rate, in particular in 3-D cases.

7. PROGRAM STRUCTURE AND TIME PARTITIONING TECHNIQUES

7.1 Flowchart of the explicit treatment of nonlinear transient dynamics

According to the preceding sections, the logical sequence for the explicit integration of the equations of motion is easily set up (Table III). Recording that motion is linearized within each time step, an important notion for the overall understanding of the explicit integration procedures, step 3 to 7 of Table III are straightforward. Natural boundary conditions are, as is usual in the finite element method, easily casted in the procedure (step 6). Energy balance, a useful tool for the a posteriori checking of computational results, is also readily casted within the procedure. The evaluation of maximum mesh frequencies is needed for the computation of the allowable time step size for the next cycle.

Essential boundary conditions could be satisfied by merely re-imposing values at the end of each step according to the imposed values or, if a more refined procedure is desired, in particular if reaction forces are to be evaluated, a specific approach based on the concept of Lagrange multipliers is proposed in the next section.

Starting conditions are rather obvious though it is important for perfect synchronization to consider non-zero external forces or internal forces generated at time t_o by an eventual out-of-equilibrium stress field. These can produce a non-zero initial acceleration field which affects the velocity in the first step. An initial velocity should thus be referred to time t_o^- and corrected for the first step according to the normal integration formula for velocities. This is automatically achieved by the insertion of step 2. Eventual non-zero initial displacements should be tackled with some caution if they are to be considered as stress generating in that the procedures shown here do not cover quasi-static behaviour. If they do not produce stress they are merely to be viewed as a change in original configuration and can as well be defined at the mesh generation stage.

The main practical drawback of the flowchart proposed in Table III is the use of a time marching algorithm with spatially constant time step. Indeed this can become an unacceptable limitation in applications involving, for instance, local mesh refinements in that the allowable size of the time step which applies to the whole mesh may drop dramatically and conduce to unfeasibility of the actual computation then required. Such difficulties can be alleviated by the use of so-called time partitioning techniques and will be examined in some detail in Section 7.3.

7.2 The treatment of essential boundary conditions

The following condition is considered:

$$\underline{C}(t)\underline{v}(t) = 0 \tag{71}$$

where $\underline{C}$ is a matrix of coefficients.

A condition on displacements reduces essentially to the same form. Supposing that a configuration n has been reached, the velocity corresponding to this configuration is not known yet. The following condition should hold, C^n stays for $C(t^n)$:

$$\underline{C}^n\underline{v}^n = 0 \tag{72}$$

In the central difference scheme (15), the velocity is expressed as:

$$\underline{v}^n = \underline{v}^{n-1} + \frac{\Delta t}{2}(\underline{\gamma}^{n-1} + \underline{\gamma}^n) \tag{73}$$

where $\underline{\gamma}$ stands for the acceleration vector $\ddot{\underline{u}}$.

Therefore, condition (72) can be rewritten as:

$$\underline{C}^n \underline{v}^{n-1} + \frac{\Delta t}{2} \underline{C}^n \underline{\gamma}^{n-1} + \frac{\Delta t}{2} \underline{C}^n \underline{\gamma}^n = 0 \tag{74}$$

and finally condition (72) corresponds to

$$\underline{C}^n \underline{\gamma}^n = \underline{S}^n \tag{75}$$

where $\underline{S}^n = - \frac{2}{\Delta t} \underline{C}^n \left[\underline{v}^{n-1} + \frac{\Delta t}{2} \underline{\gamma}^{n-1} \right].$

Furthermore, $\underline{\gamma}^n$ is present in the equilibrium equation

$$\underline{M}^n \underline{\gamma}^n = \underline{F}^n \tag{76}$$

where $\underline{F}^n = \underline{f}^{ext\,n} - \underline{f}^{int\,n}$.

The constraint defined by Eq.(75) is introduced in (76) by means of Lagrange multipliers. The reaction forces corresponding to the constraint are expressed as:

$$\underline{F}_c^n = \underline{C}^{nT} \underline{\lambda}^n \tag{77}$$

where $\underline{\lambda}$ are the Lagrange multipliers.

In order to obtain an expression for the Lagrange multipliers, the modified equilibrium equation

$$\underline{M}^n \underline{\gamma}^n = \underline{F}^n + \underline{F}_c^n \tag{78}$$

is multiplied by $\underline{C}^n$ and the reaction forces substituted in terms of Lagrange multipliers:

$$\underline{C}^n \underline{\gamma}^n = \underline{C}^n \underline{M}^{-1\,n} \underline{F}^n + \underline{C}^n \underline{M}^{-1\,n} \underline{C}^{T\,n} \underline{\gamma}^n \tag{79}$$

and applying the constraint (75), the following expression permitting to find the Lagrange multipliers is obtained:

$$\underline{C}^n \underline{M}^{-1\,n} \underline{C}^{T\,n} \underline{\lambda}^n = \underline{S}^n - \underline{C}^n \underline{M}^{-1\,n} \underline{F}^n \tag{80}$$

Associated reaction forces to include in the equilibrium equations are then obtained via Eq.(77).

The solution of a small system involving only the degrees of freedom tied by the constraint does thus permit a general treatment of the constraint. Within the flowchart presented in Table III, this fits into step 6 and modifies in no way the general procedure proposed.

7.3 Time partitioning techniques

In order to increase the time step imposed by the conditionally stable character of explicit methods, it is often proposed to use partitioning techniques in particular explicit-implicit partitions [6]. The idea is to treat elements leading, in an explicit environment, to too severe stability restrictions, in an implicit fashion recovering thereby unconditional stability for these elements. While this may apparently be thought of a convenient method for alleviating the uneconomical consequences of stringent localized stability conditions, a number of points should, however, be noted:

- the class of problems of interest in nonlinear transient dynamics is essentially suited for an explicit treatment;
- in particular local phenomena including wave propagation and strong nonlinearities (geometrical and material) are best treated in an explicit environment;
- implicit treatment needs computational tools which are not at disposal in an exclusively explicit programme;
- programme complexity augments significantly in the presence of explicit-implicit partitioning techniques.

Therefore, preference is given here to explicit-explicit partitioning [4].

The stability conditions (see Section 3.2) associate a time step limitation to the explicit computation of velocity components. This integration process corresponds in discretized form to the satisfaction of conservation of momentum and it is essential for accuracy reasons and also for coherence in the assumptions to proceed with some care. In order to understand how partitioning will be achieved, it should first be noted that the computation of velocity components pertains to nodes and not to elements. On the other hand, stability conditions are usually derived on an element basis. The integration producing a new velocity at a defined node depends thus on the stability restrictions found in neighbouring elements and more precisely the time step which can be used for velocity update must remain bounded by the most stringent condition found. Applying this thinking to the entire structure, a critical time step can be associated with each node. Computing the velocity according to this nodal map of time steps would result straightforwardly in a stable and, therefore, accurate solution if two main difficulties were not encountered:

- the update of velocity requires the knowledge of the current state of stress in adjacent elements (even those, for instance, where no stability restriction is found);
- some sort of synchronization is needed in order to establish a useful solution scheme.

In order to cope with this situation a further time step map is constructed relative now to the elements in order to be used for configuration update. An economically less attractive solution could consist in computing always the configuration of the entire structure. At the same time when the maps are created they are modified according to a binary pattern in order to permit synchronization. Programming details are beyond the scope of this paper and it is simply noted here that the original programme structure of spatially constant time integration presented in Table III can indeed be easily adapted in order to cope with fully automatic time partitioning.

REFERENCES

1. Key, S.W.: A finite element procedure for large deformation dynamic response of axisymmetric solids, Comp. Meths. Appl. Mech. Engng., 4 (1974), 195-218.
2. Key, S.W.: Transient response by time integration: review of implicit and explicit operators, in Advanced Structural Dynamics (ed. J. Donea), Applied Science, London, 1980, 71-95.
3. Key, S.W. and Krieg, R.D.: On the numerical implementation of inelastic time dependent and time independent finite strain constitutive equations in structural mechanics, Comp. Meths. Appl. Mech. Engng., 33 (1982), 439-452.
4. Key, S.W.: Improvements in transient dynamic time integration with application to spent nuclear fuel shipping cask impact analyses, Proc. of the Eur.-US Symposium on Finite Element Methods for Nonlinear Problems, Trondheim, August 12-16, 1985 (eds. P.G. Bergan, K.J. Bathe and W. Wunderlich), Springer-Verlag, Berlin 1986.
5. Zienkiewicz, O.C.: The Finite Element Method, Mc-Graw Hill, Maidenhead 1977.
6. Belytschko, T. and Hughes, T.J.R. (eds.): Computational Methods for Transient Analysis, Vol.1, in: Computational Methods in Mechanics; Mechanics and Mathematical Methods (gen. ed. J.D. Achenbach), North-Holland, Amsterdam 1983.
7. Hughes, T.J.R.: Numerical implementation of constitutive models: rate-independent deviatoric plasticity, in: Proc. of the Workshop on the Theoretical Foundation for Large-Scale Computations of Nonlinear Material Behaviour, Northwestern University, Evanston, Illinois, October 24-26, 1983 (eds. S. Nemat-Nasser, R.J. Asaro and G.A. Hegemier), Martinus Nijhoff Publishers, Dordrecht 1984, 29-57.
8. Halleux, J.P. and Donea, J.: a discussion of Cauchy stress formulations for large strain analysis, in: Proc. of the Eur.-US Symposium on Finite Element Methods for Nonlinear Problems, Trondheim, August 12-16, 1985 (eds. P.G. Bergan, K.J. Bathe and W. Wunderlich), Springer-Verlag, Berlin 1986.
9. Truesdell, C. and Toupin, R.A.: The Classical Field Theories, Vol.III/1, in: Encyclopedia of Physics (ed. S. Flügge), Springer-Verlag, Berlin 1960.

Truesdell, C. and Noll, W.: The Nonlinear Field Theories of Mecha-

nics, Vol.III/3, in: Encyclopedia of Physics (ed. S. Flügge), Springer-Verlag, Berlin 1965.
10. Dienes, J.K.: On the analysis of rotation and stress rate in deforming bodies, Acta Mechanica, 32 (1979), 217-232.
11. Green, A.E. and McInnis, B.C.: Generalized hypoelasticity, in: Proc. Roy. Soc. Edinburgh A57, Part III (1967), 220-230.

TABLE 1 (from [7]) - Summary of elastic-plastic constitutive theory with Von Mises yield surface, associative flow rule, and linear combination of isotropic and kinematic hardening or softening

Constitutive equation: $\dot{\underline{\sigma}} = \underline{C} \cdot (\dot{\underline{\varepsilon}} - \dot{\underline{\varepsilon}}^p)$

Evolution equation for the radius of the yield surface:

$$\dot{R} = \frac{\sqrt{2}}{\sqrt{3}} \beta H \dot{\bar{\varepsilon}}^p$$

Evolution equation for the back stress:

$$\dot{\underline{\alpha}} = \frac{2}{3} (1-\beta) H \dot{\underline{\varepsilon}}^p$$

Flow rule: $\dot{\underline{\varepsilon}}^p = \begin{cases} 0 & \text{if (E)} \\ \dot{\lambda}\underline{N} & \text{if (P)} \end{cases}$

Consistency condition: $\dot{\lambda} = \frac{1}{2\mu} \frac{1}{\left(1 + \frac{H}{3\mu}\right)} \underline{N} \cdot \dot{\underline{\sigma}}^{tr} = \frac{1}{\left(1 + \frac{H}{3\mu}\right)} \underline{N} \cdot \dot{\underline{\varepsilon}}$

Trial rate of stress: $\dot{\underline{\sigma}}^{tr} = \underline{C} \cdot \dot{\underline{\varepsilon}}$

Definition of an elastic process (E):

$$\left.\begin{array}{lll} f(\underline{\xi}) < 0, & \text{or} & \\ f(\underline{\xi}) = 0 & \text{and} & \underline{N} \cdot \dot{\underline{\sigma}}^{tr} \leq 0 \end{array}\right\}$$

Definition of a plastic process (P):

$$f(\underline{\xi}) = 0 \quad \text{and} \quad \underline{N} \cdot \dot{\underline{\sigma}}^{tr} > 0$$

Unit normal: $\underline{N} = \underline{\xi}'/R$ where $\underline{\xi} = \underline{\sigma} - \underline{\alpha}$

(the prime denotes deviatoric part)

$\underline{N}$ being a unit normal, $\dot{\lambda} = \sqrt{\frac{3}{2}} \dot{\Lambda}$

TABLE II (from [7]) - Krieg and Key's radial-return algorithm for the elastic-plastic case with Von Mises yield surface, associative flow rule and linear combination of isotropic and kinematic hardening or softening

Step 1: Calculate the trial stresses:

$$\underline{\sigma}_{n+1}^{tr} = \underline{\sigma}_n + \underline{C} \cdot \Delta\underline{\varepsilon}$$

$$\underline{\xi}_{n+1}^{tr} = \underline{\sigma}_{n+1}^{tr} - \underline{\alpha}_n$$

Step 2: Calculate the mean component of $\underline{\xi}_{n+1}^{tr}$:

$$\xi = \frac{1}{3} \text{ trace } \underline{\xi}_{n+1}^{tr}$$

Step 3: Calculate the deviatoric part of $\underline{\xi}_{n+1}^{tr}$:

$$\underline{\eta} = \underline{\xi}_{n+1}^{tr} - \xi \, \underline{I}$$

Step 4: $A = \underline{\eta} \cdot \underline{\eta}$

Step 5: If $A \leqq R_n^2$ (i.e. elastic process), then:

$$\underline{\sigma}_{n+1} = \underline{\sigma}_{n+1}^{tr}, \quad R_{n+1} = R_n \quad \text{and} \quad \underline{\alpha}_{n+1} = \underline{\alpha}_n$$

Otherwise (i.e. plastic process), continue.

Step 6: Calculate the approximation to the normal:

$$\underline{N} = \underline{\eta}/|\underline{\eta}|$$

Step 7: Calculate $\tilde{\Lambda}$:

$$\tilde{\Lambda} = \frac{1}{2\mu} \frac{1}{\left(1 + \frac{H}{3\mu}\right)} \left(|\underline{\eta}| - R_n\right)$$

Step 8: Update:

$$\underline{\sigma}_{n+1} = \underline{\sigma}_{n+1}^{tr} - 2\mu \, \tilde{\Lambda} \, \underline{N}$$

$$R_{n+1} = R_n + \frac{2}{3} \beta \, H \, \tilde{\Lambda}$$

$$\underline{\alpha}_{n+1} = \underline{\alpha}_n + \frac{2}{3} (1-\beta) H \, \tilde{\Lambda} \, \underline{N}$$

TABLE III - Flowchart

Step 1:	Set initial conditions: $n=0$, $\Delta t=0$, $t^n=t_o$, $\underline{x}^n=\underline{x}_o$, $\underline{\sigma}^n=\underline{\sigma}_o$, $\dot{\underline{u}}^n=\dot{\underline{u}}_o$ $W^{int}=0$ (internal energy), $W^{ext}=W^{kin}$ (external and kinetic energy)
Step 2:	GO TO Step 5.
Step 3:	(note: $\underline{x}^n, \underline{\sigma}^n, \ddot{\underline{u}}^n$ and $\underline{v}^{n+\frac{1}{2}}=\dot{\underline{u}}^n+\frac{\Delta t}{2}\ddot{\underline{u}}^n$ are known quantities) $n=n+1$, $\Delta t=\Delta t^{new}$ and $t^n=t^n+\Delta t$.
Step 4:	Update configuration $\underline{x}^n=\underline{x}^{n-1}+\Delta t\ \underline{v}^{n+\frac{1}{2}}$ (loop over nodes)
Step 5:	Compute internal forces $\underline{f}^{int}=\sum_e \int_{V^e} (\underline{B})^T\underline{\sigma}^n dv$ (loop over elements) V^e and $(\underline{B})^T$ are evaluated on the new configuration (n). $\underline{\sigma}^n=\underline{\sigma}^{n-1}+\Delta\underline{\sigma}$; $\Delta\underline{\sigma}$ evaluated according to sections 5 and 6. While computing element stresses, evaluate ω^e_{max} for next time step setting Δt^{new} and add internal energy increment to W^{int}.
Step 6:	Assemble internal forces, evaluate and assemble external forces ($\underline{f}^{ext}$) and add external energy increment to W^{ext}.
Step 7:	Compute accelerations and update velocities $\ddot{\underline{u}}^n = \underline{M}^{-1}\ (\underline{f}^{ext\,n} - \underline{f}^{int\,n})$ (loop over nodes) $\underline{v}^{n+\frac{1}{2}} = \underline{v}^{n-\frac{1}{2}} + \frac{\Delta t+\Delta t^{new}}{2}\ddot{\underline{u}}^n$ (loop over nodes)
Step 8:	If no output required, GO TO Step 3.
Step 9:	Compute kinetic energy and check energy balance: $W^{ext} \cong W^{int} + W^{kin}$ Print required output.
Step 10:	If computation not ended, GO TO Step 3.

CHAPTER 3

EARTHQUAKE INPUT DEFINITION AND THE TRASMITTING BOUNDARY CONDITIONS

O. C. Zienkiewicz, N. Bicanic
University College of Swansea, Swansea, UK

F. Q. Shen
Hehai University, Nanjing, China

ABSTRACT

The specification of the earthquake input for the linear and nonlinear analysis of the structure-foundation problem is usually done by prescribing the base motion. Most commonly the actual analysis is carried out in terms of displacement relative to this base movement. If realistic boundary conditions need to be specified because of the infinite extent of the foundation, this "conventional" procedure is not applicable. An alternative is therefore suggested by specifying the motion in terms of the incoming seismic wave, leading to a logical and simple problem formulation.

1. INTRODUCTION

Seismic analysis involving the assumption of a rigid foundation base moving with a prescribed motion $u_b(t)$ is well understood. The problem can be formulated either in terms of total displacementsor in terms of displacements relative to the rigid foundation base.

The formulation of the problem requires the definition of the earthquake signal, defined either in terms of displacements or accelerations, with the latter being the most commonly adopted approach. It should be noted however that this traditional modelling process assumes the t o t a l motion of the foundation base to be given, i.e. total displacements or total accelerations.

This approach cannot readily be applied in cases where no distinct base foundation or base rock exists (Fig.1), and when the truncation of the mathematical model has to be chosen arbitrarily. The truncation in this context refers here to any model boundary (the bottom and lateral model boundaries for the familiar soil structure interaction problem in two dimensions , for example). To account for such a case correctly it is necessary to incorporate the appropriate radiation boundary conditions, such that o u t g o i n g wave (the one travelling in the direction of the boundary outer normal) is not reflected from this arbitrary boundary. Clearly, if the total motion were to be defined for such a boundary (like in the traditional rigid base approach), no radiation boundary conditions could be imposed at the same time. It seems therefore appropriate to formulate the problem in terms of the i n c o m i n g and the o u t g o i n g waves so that the radiation boundary conditions for the outgoing waves can be imposed. For that approach to be viable, the i n c o m i n g wave on the model truncation boundary needs to be known. In the following sections it will be assumed for simplicity that the incoming seismic wave is represented by a vertically propagating plane wave (compression or shear).

In the case of the dynamic loading applied to the surface the truncation boundary is only required to transmit the outgoing waves. However in the case of the earthquake loading the truncation boundary is more complex, as it has to allow the incoming seismic wave to "enter" the model, as well as to ensure that the outgoing waves are transmitted.

2. DEFINITION OF THE INCOMING SEISMIC WAVE

The seismic signal is usually measured at or near the free surface, and it represents the modification of the original seismic wave which initiated at the earthquake source, caused by passing through different material zones and involving number of reflections and refractions at the interfaces between zones or layers of different material.

The geological conditions at the site will very often be such that the so called "bedrock" exists,i.e a zone of significantly more rigid material is situated underneath the softer soil layers. Any incoming seismic wave passing from the bedrock to the softer soil layers will amplify depending on the material properties of both bedrock and soil layers. The significant consequence of the presence of the bedrock lies in the fact that any of the reflected waves (either from interfaces between the soil layers or from the free surface) are practically

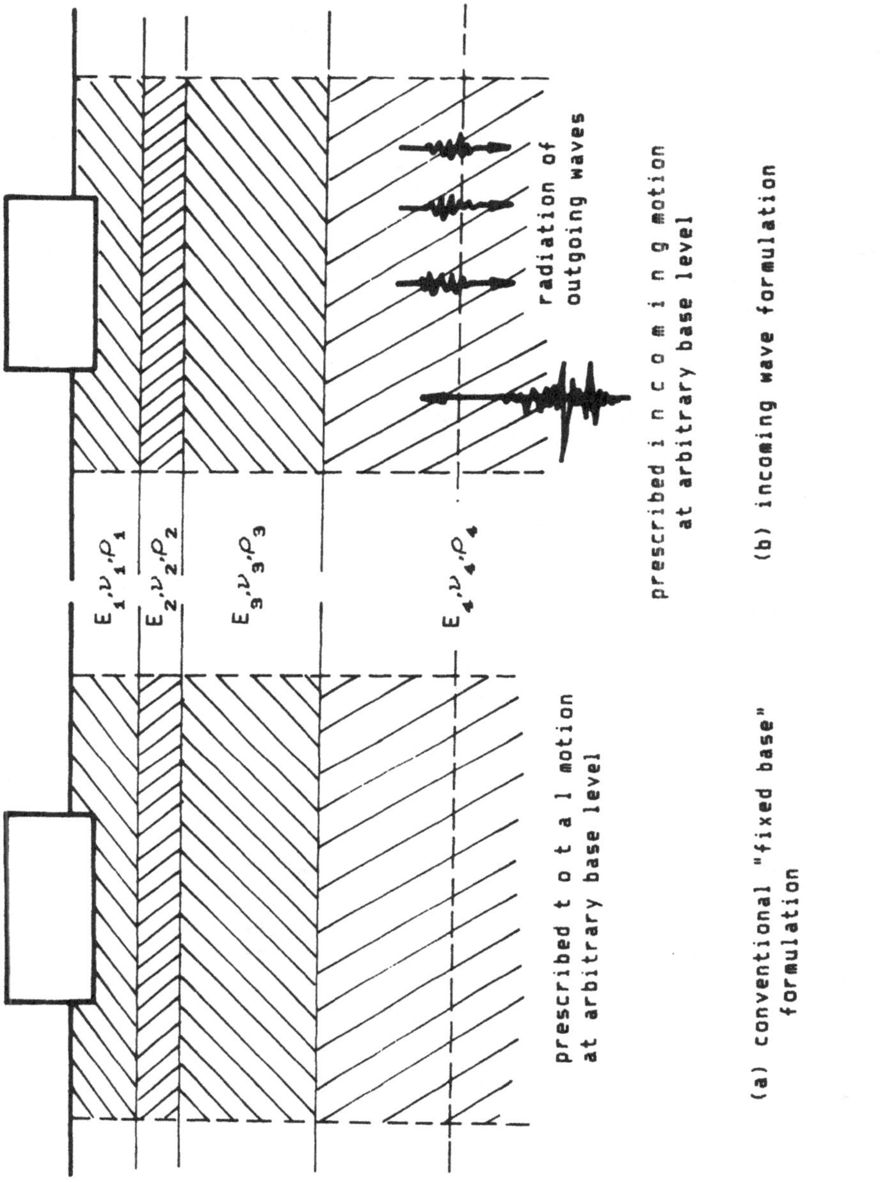

Fig 1 - The structure-foundation problem

trapped inside the soft soil layers,as only a small fraction can be transmitted back to the bedrock on the interface between the softer soil layers and the bedrock. As the bedrock is significantly more rigid, the transmitted wave is smaller than the one reflected back towards the soil surface. In such a case, the traditional fixed base approach is valid and no transmitting boundary conditions need be imposed on the bedrock level as practically no waves get transmitted into the bedrock.

The need for an arbitrary model truncation emerges in the cases where no distinct base rock exists, or when the exent of the softer soil layers is so great,that it would be prohibitive to include the whole zone into the mathematical model. Such a situation may also arise when the nonlinear material behaviour can be expected only near the surface and deeper layers (with material properties still far from bedrock like characteristics) are expected to remain elastic .

To model such a case correctly it is necessary to reconstruct the incoming seismic wave at the model truncation boundary. In the simplest case of a one dimensional elastic, homogeneous, isotropic wave propagation problem involving the free surface, it is very well known that the free surface displacement wave equals the double of the incoming displacement wave. Here the incoming signal can easily be extracted from the recorded total signal on the undisturbed surface. Even in the case of the elastic non homogeneous domain the incoming signal can again be extracted from the total signal recorded on the surface. Therefore, in the following ,it will be assumed that the incoming wave (displacement, velocity, acceleration) is known at a position corresponding to the model truncation boundary, and that outside this homogeneous elastic conditions pertain.

3. BASIC CONCEPTS IN A ONE DIMENSIONAL CASE

The basic idea can be best illustrated on a one dimensional example (Fig 2) where it is assumed that below certain level (denoted here with line AA) the conditions of elasticity and homogenuity exist. Above the level AA, the material may well be nonlinear and nonhomogeneous. The earthquake input is defined by an incoming wave in the homogeneous, elastic medium below AA, and an arbitrary model truncation level BB is set at x = 0.

The incoming wave may be defined by the displacement

$$u_I(x,t) = u_I(x - ct) \tag{1}$$

which at level BB takes the value

$$u_{IO}(t) = u_I(0,t) \tag{2}$$

As it will be seen later, the specification of the above incoming motion is the only requirement for the appropriate analysis, and replaces the "conventional" base motion definition.

3.1. Displacement solution splitting

A convenient way of splitting the total displacement u is now made such that

$$u\ (x,t) = u_{IO}\ (t) + u_{R}\ (x,t) \tag{3}$$

where $u_{IO}(t)$ represents the history of the incoming wave at the truncation level BB (x =0) , and $u_R(x,t)$ is the displacement relative to the incoming displacement at the truncation level. Note that $u_R(x,t)$ contains not only outgoing waves but waves travelling both upwards and downwards, including all multiple reflections on layer interfaces, for every point $x > 0$. However, at $x = 0$, $u(0,t)$ represents only the outgoing wave, and it is this wave that needs to be transmitted through the truncation level. It is important to point out that in the split of equation (3) u_{IO} does not vary with x.

The equation of motion in the domain of interest (in the absence of internal damping) is

$$\frac{\partial \sigma}{\partial x} - \rho \frac{\partial^2 u}{\partial t^2} = 0 \tag{4}$$

with

$$\sigma = D \frac{\partial u}{\partial x} \tag{5}$$

for simple linear elasticity, where D denotes the appropriate elastic modulus (E in the case of pressure waves and G in the case of shear waves) .

For the nonlinear case a tangential material modulus $D(\sigma)$ relating the increments of stresses and strains can be defined by

$$d\sigma = D(\sigma)\ d\left(\frac{\partial u}{\partial x}\right) \tag{6}$$

Substituting the split of equation (3) in the equation (4), the modified equation of motion in terms of u_R is obtained

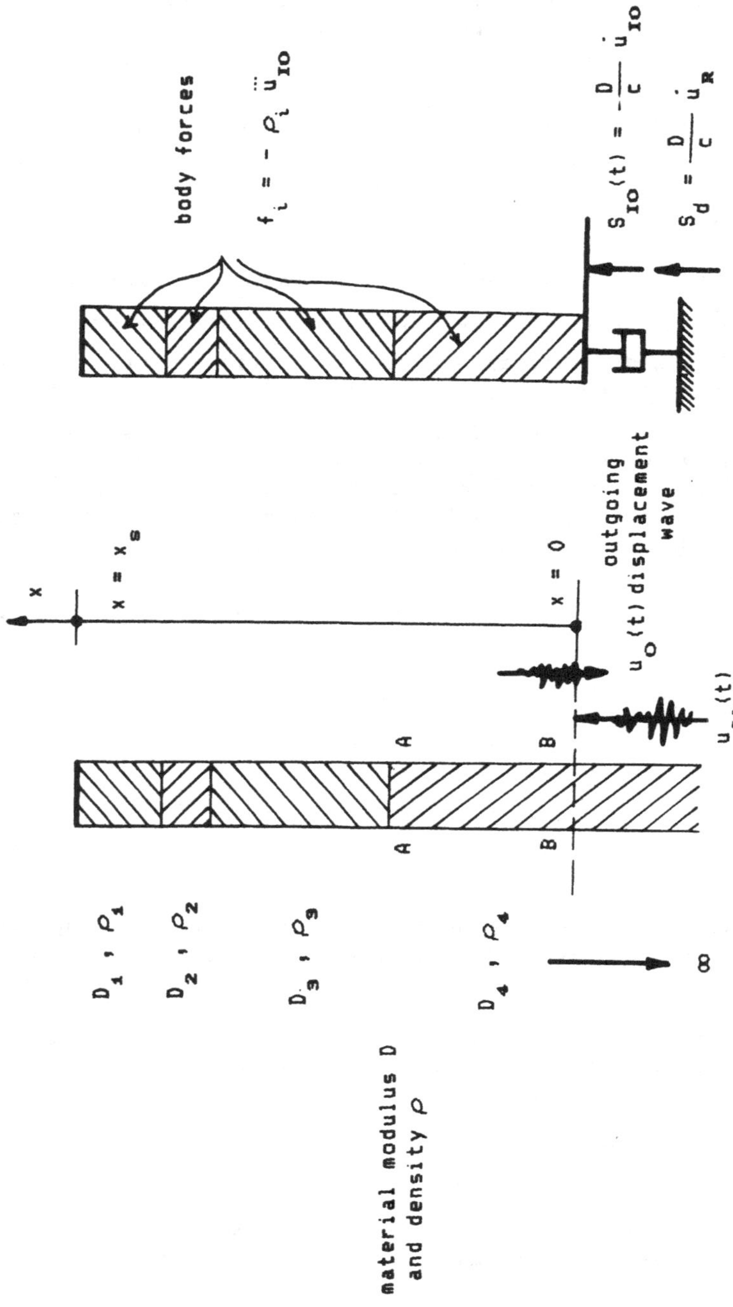

Fig 2 - One dimensional incoming seismic wave formulation

$$\frac{\partial \sigma}{\partial x} - \rho \frac{\partial^2 u_R}{\partial t^2} = \rho \frac{\partial^2 u_{IO}}{\partial t} \tag{7}$$

as u_{IO} does not vary with x .

The format of the above equation is the same as that of the traditional ones involving the rigid base assumption, the difference being that the right hand side contains now the incoming, rather then the total acceleration at the model truncation level. However, unlike the traditional fixed base formulation, where the solution unknown represents the displacement relative to the total displacement of the model base, here u_R represents for any point in the domain (x > 0) the displacement relative to the known incoming displacement wave at the truncation level (x = 0), hence the total displacement at the model base u(0,t) is not equal to u_{IO} .

3.2. Boundary conditions

Unlike the traditional rigid base formulation,where the boundary conditions present no difficulty , but do not allow the transmission of the outgoing waves, now the boundary conditions for the solution variable u_R in equation (7) have to defined in such a way to allow this. It can be noted that the solution variable u_R in equation (7) is not equal to the outgoing wave, except at the truncated boundary.

Boundary conditions for u_R need to be defined $x = x_S$, corresponding to the free surface , and at the truncation level x = 0.

The condition on the free surface ($x = x_S$) is simply

$$\sigma\Big|_{x_S} = D\,\frac{\partial u}{\partial x}\Big|_{x_S} = 0$$

or

$$\frac{\partial u}{\partial x}\Big|_{x_S} = \frac{\partial(u_{IO} + u_R)}{\partial x}\Big|_{x_S} = \frac{\partial u_R}{\partial x}\Big|_{x_S} = 0 \tag{8}$$

so the boundary condition for u_R on the free surface is simply the specification of zero traction in terms of this variable.

On the bottom truncation boundary (x = 0) , the appropriate boundary condition corresponds to the transmission of the outgoing wave.

In the purely elastic homogeneous domain (below level AA) the equation of motion in terms of total displacement is written as

$$D \frac{\partial^2 u}{\partial x^2} - \rho \frac{\partial^2 u}{\partial t^2} = 0 \tag{9}$$

for which case the well known wave solution is of the form

$$u = u_I (x - ct) + u_O (x + ct) \tag{10}$$

where the wave propagation speed is $c = \left(\frac{D}{\rho} \right)^{1/2}$

In the above u_O represents the outgoing wave, for which the well known radiation condition [1,2] implies that

$$\left. \frac{\partial u_O}{\partial x} \right|_0 = \frac{1}{c} \left. \frac{\partial u_O}{\partial t} \right|_0 \tag{11}$$

However, as the equation of motion (7) is not written in terms of u_O, but in terms of u_R (which, except on the truncation boundary is n o t equal to the outgoing wave), the boundary conditions for u_R on the truncation boundary have to be defined.

From the nature of the adopted displacement split of equation (3) it follows that $u_O = u_R$ at $x = 0$, and hence

$$\left. \frac{\partial u_O}{\partial t} \right|_0 = \left. \frac{\partial u_R}{\partial t} \right|_0 \tag{12}$$

and the right hand side of the radiation condition for outgoing wave u_O (11) is now redefined in terms of u_R .

From the equations (3) and (10) it follows that in the homogeneous region bellow AA

$$u_I (x,t) + u_O (x,t) = u_{IO} (t) + u_R (x,t)$$

and consequently

$$u_O (x,t) = u_{IO} (t) + u_R (x,t) - u_I (x,t) \tag{13}$$

Furthermore, as the incoming wave u_I can be written in a form

$$u_I = u_I\,(x - ct) \tag{14}$$

it follows that

$$\frac{\partial u_I}{\partial x} = u_I' \,, \qquad \frac{\partial u_I}{\partial t} = -c\,u_I' \,, \qquad \frac{\partial u_I}{\partial x} = -\frac{1}{c}\,\frac{\partial u_I}{\partial t} \tag{15}$$

and

$$\left.\frac{\partial u_I}{\partial x}\right|_0 = -\frac{1}{c}\left.\frac{\partial u_I}{\partial t}\right|_0 = -\frac{1}{c}\,\frac{\partial u_{IO}}{\partial t} \tag{16}$$

as u_{IO} is equal to u_I for $x = 0$.

Differentiating equation (13) with respect to x and noting that $\frac{\partial u_{IO}}{\partial t} \equiv 0$, as u_{IO} does not vary with x, the left hand side of the radiation condition (11) can be redefined in terms of u_R as

$$\left.\frac{\partial u_O}{\partial x}\right|_0 = \left.\frac{\partial u_R}{\partial x}\right|_0 - \left.\frac{\partial u_I}{\partial x}\right|_0 \tag{17}$$

On substituting (16) this gives

$$\left.\frac{\partial u_O}{\partial x}\right|_0 = \left.\frac{\partial u_R}{\partial x}\right|_0 + \frac{1}{c}\,\frac{\partial u_{IO}}{\partial t} \tag{18}$$

Substituting (12) and (18) into (11) leads to the radiation boundary condition for the outgoing wave u_O written in terms of the solution variable u_R.

$$\left.\frac{\partial u_R}{\partial x}\right|_0 + \frac{1}{c}\,\frac{\partial u_{IO}}{\partial t} = \frac{1}{c}\left.\frac{\partial u_R}{\partial t}\right|_0 \tag{19}$$

Alternatively this can be written in terms of the boundary traction S on that boundary, as

$$S = D \left.\frac{\partial u_R}{\partial x}\right|_0 = \frac{D}{c} \left(\left.\frac{\partial u_R}{\partial t}\right|_0 - \frac{\partial u_{IO}}{\partial t} \right) \tag{20}$$

As u_{IO} is a known function of time, the boundary traction S can be interpreted physically as a superposition of the "standard" viscous damper and the varying prescribed force S_{IO} on the boundary

$$S = \frac{D}{c} \left.\frac{\partial u_R}{\partial t}\right|_0 - S_{IO} \tag{21}$$

where S_{IO} is defined from the known incoming wave signal as

$$S_{IO} = \frac{D}{c} \frac{\partial u_{IO}}{\partial t} \tag{22}$$

3.3. Discretixed equations of motion

The finite element discretization [2] of the equation of motion (4,7) leads to the form

$$\underset{\sim}{M}\, \underset{\sim R}{u} + \underset{\sim}{K}\, \underset{\sim R}{u} + [\, \underset{\sim}{P}\, \underset{\sim R}{u} \,]_\Gamma = - \underset{\sim}{M}\, \underset{\sim IO}{u} + [\, \underset{\sim}{S} \,]_\Gamma \tag{23}$$

where the similarity with the conventional rigid base approach is again noted, now however with two additional terms contributed by the boundary conditions, i.e.

$$\underset{\sim}{P} = \int_\Gamma \underset{\sim}{N}^T \left(\frac{D}{c}\right) \underset{\sim}{N}\, d\Gamma \qquad \text{and} \qquad \underset{\sim}{S} = \int_\Gamma \underset{\sim}{N}^T \left(\frac{D}{c}\right) \dot{\underset{\sim IO}{u}}\, d\Gamma \tag{24}$$

Further, the right hand side involves now the incoming acceleration history as opposed to the total base acceleration in the conventional format. The truncation boundary degrees of freedom for u_R are unconstrained and the radiation boundary condition for u_O is ensured by employing the viscous damper and the time varying traction as indicated in equation (24). For comparison, it may be noted that the boundary degrees of freedom for the relative displacement in the conventional approach are constrained to zero.

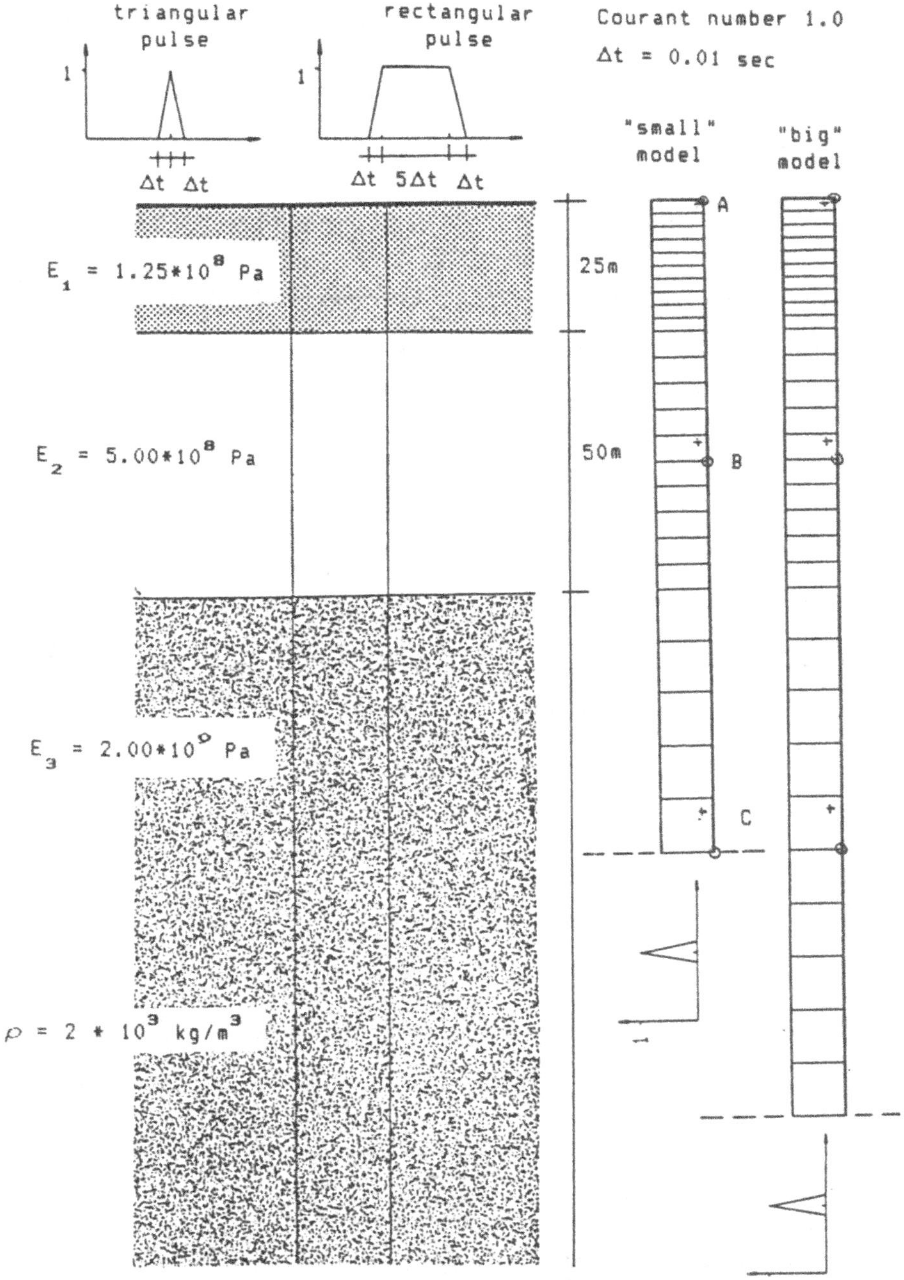

Fig 3 - Layered foundation incoming pulse wave model problem in 1D with location of control nodal and stress sampling points

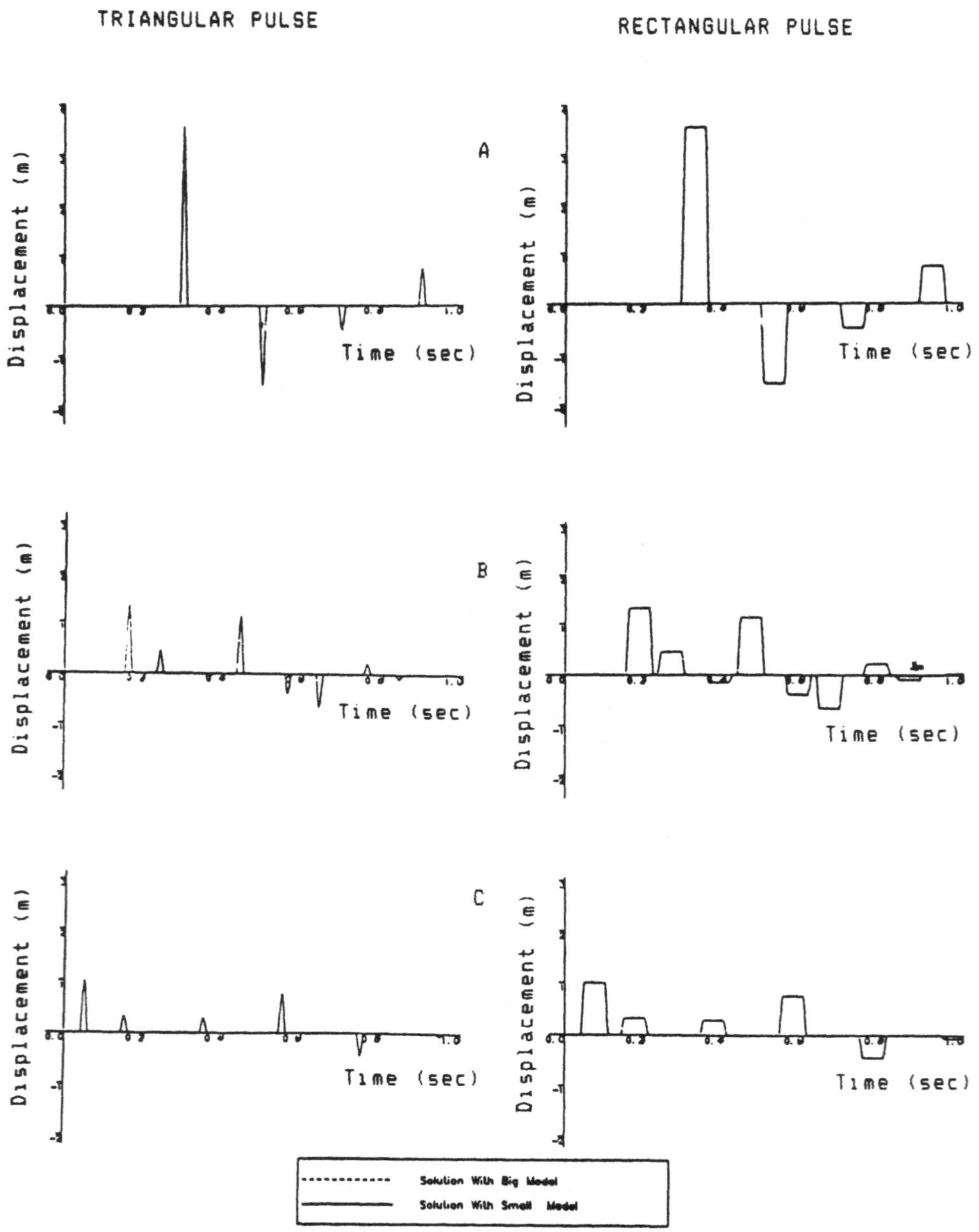

Fig 4 - Displacement histories for control points A, B, C for two incoming wave pulse wave cases and two different model truncation levels (when only one curve shown the results for "small" and "big" model identical)

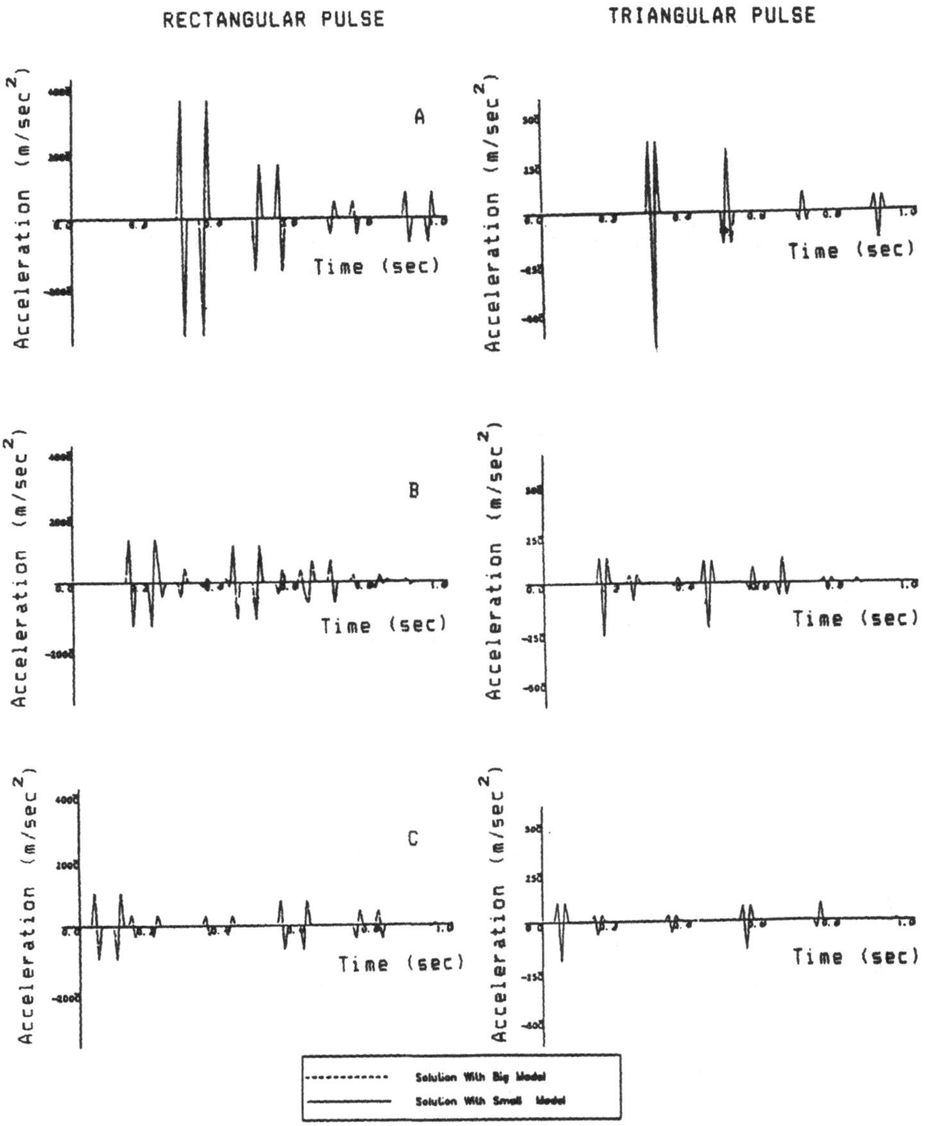

Fig 5 - Acceleration histories for control points A, B, C for two incoming pulse wave cases and two different model truncation levels (when only one curve shown, results the results for "small" and "big" model identical)

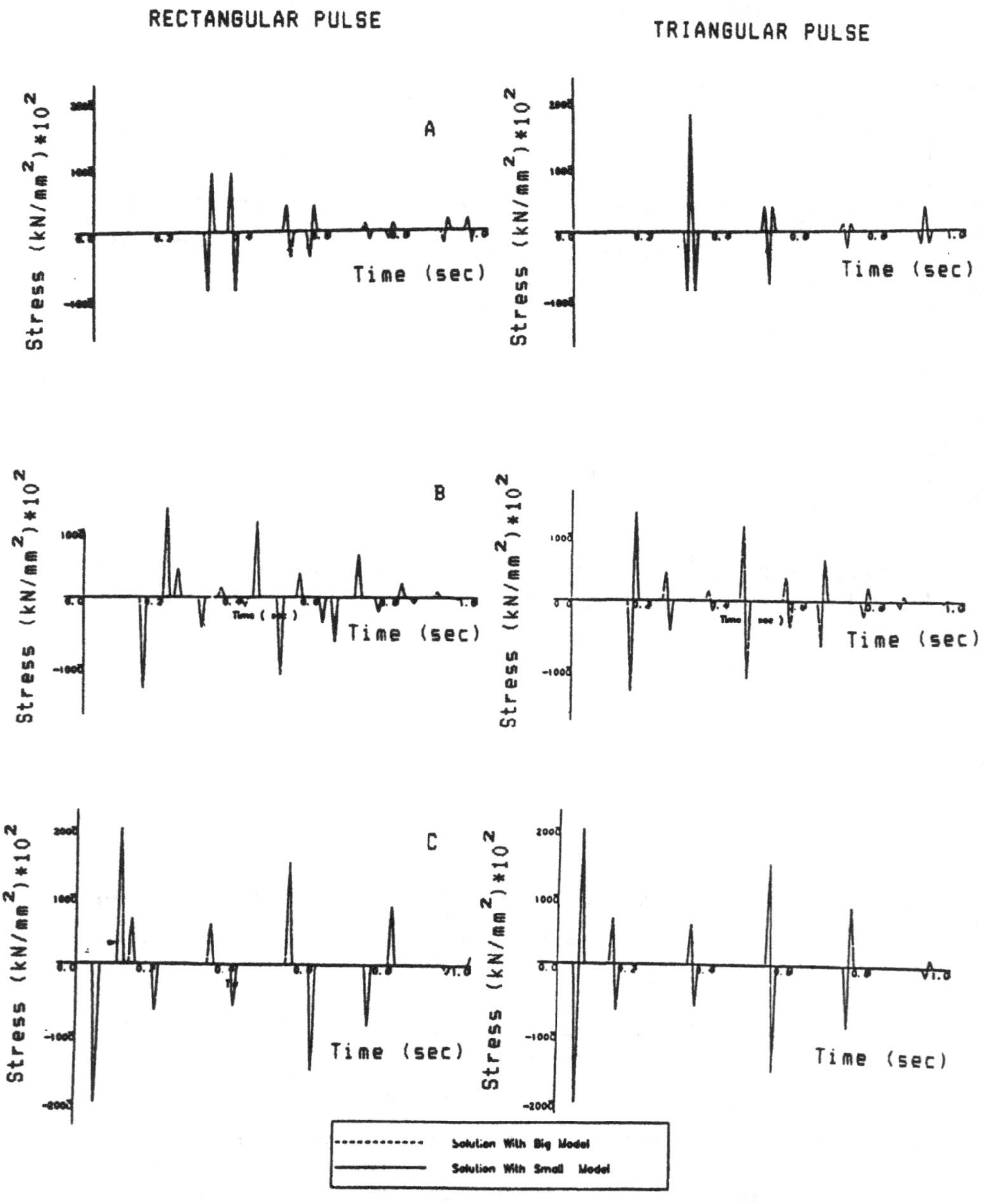

Fig 6 - Stress histories for control sampling points A,B,C for two incoming pulse wave cases and two different model truncation levels (when only one curve shown, results the results for "small" and "big" model identical)

The solution procedures to solve the discretized equations of motion can be either in explicit, implicit or mixed format and clearly any of numerous numerical schemes can be adopted. Instead of the viscous damper in equation (21) any of the equivalent transmitting boundary idealisations (Generalised Smith Boundary, Paraxial transmitting boundary) can be used [3,4,6].

3.4. Model problem

To illustrate the procedure outlined in the previous sections, a one dimensional problem involving non homogeneous soil layers (Fig.3) and a vertically propagating incoming wave is considered. Triangular and rectangular pulse incoming pressure wave are analysed for two models corresponding to different truncation levels within the homogeneous layer extending to infinity. Fully explicit integration scheme has been adopted with Courant number equal 1.0 for the homogeneous soil layer. Standard viscous damper is employed at the model truncation boundary. The near equality of the displacement, acceleration and stress histories (Figs. 4 ,5 and 6) for three control points A, B and C indicate the validity of the solutions - the only difference in the results is due to numerical dispersion in different meshes used.

4. TWO DIMENSIONAL FORMULATION - STRUCTURE ON THE HORIZONTALLY LAYERED FOUNDATION SUBJECT TO THE PLANE VERTICALLY PROPAGATING EARTHQUAKE INPUT SIGNAL

Considering the same foundation domain with a vertically propagating incoming seismic input as in the preceding sections but now including the perturbation on the surface in the form of the structure, fill or excavation, the simple one dimesional behaviour is no longer maintained. The perturbation of the surface gives rise to waves which are no longer reflected in the vertical direction only - and the need for the transmitting lateral boundaries becomes obvious, in a similar way as it would be required for a surface force excitation.

Various analysis procedures are possible and for all of these it is clear (Fig 7) that the solution for the lateral boundary points should in the limit approach the results of the one dimensional "free-field column", as the lateral model truncation boundaries Γ_{SL} and Γ_{SR} are moved away from the perturbation. For consistency purposes, such a comparative analysis should be carried out using similar finite elements and similar mesh subdivision to minimise the differences which are due to discretization errors.

The equations of motion are again written in terms of u_R, the displacement relative to the incoming seismic displacement at the model truncation boundary, and the discretized form is very

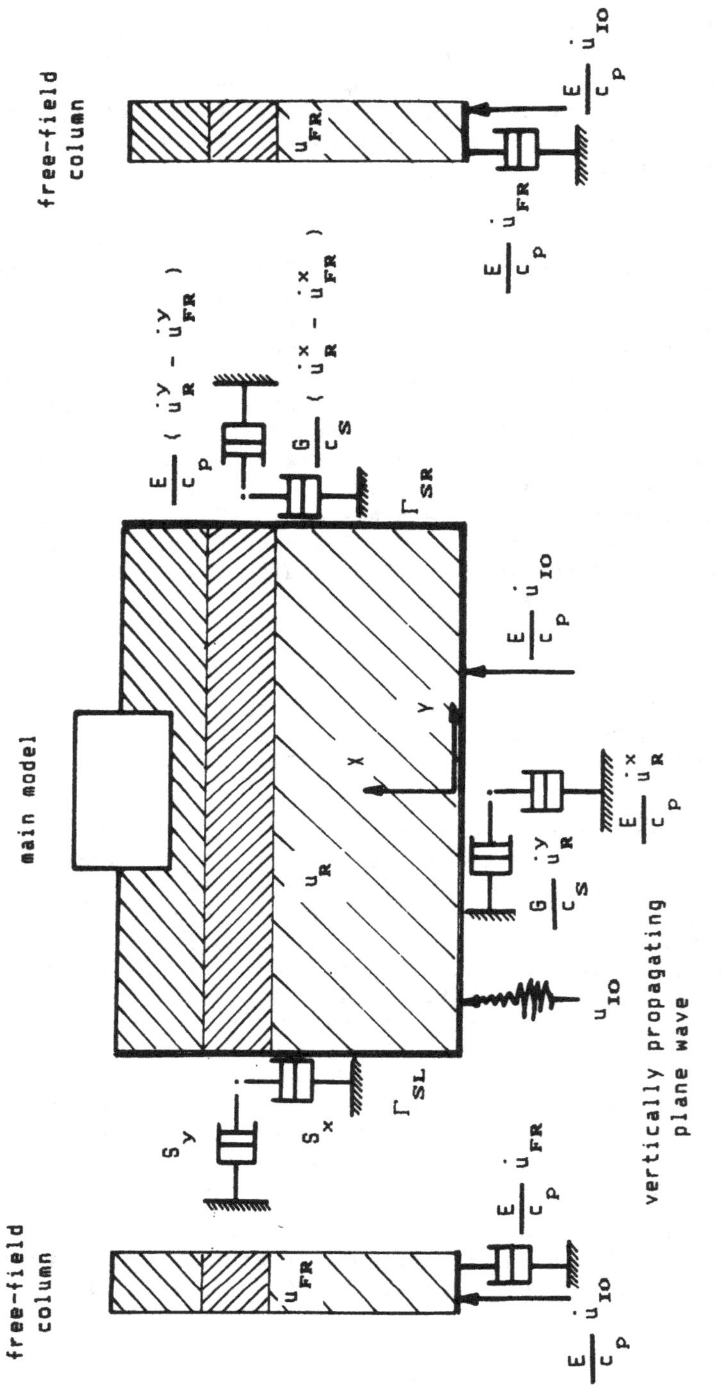

Fig 7 - Two dimensional incoming seismic wave formulation involving both bottom and lateral model boundaries

similar to the equation (23). The bottom model truncation boundary is treated in the same way as in the one dimensional case and the only additional problem arises from the treatment of the appropriate lateral boundary conditions. There, it is necessarry to allow the radiation of any waves caused by the surface perturbation, i.e. the difference between the two dimensional solution u_R and the one dimensional free-field solution u_{FR} ,which is here denoted u_{TR} . Thus

$$\underset{\sim}{u}_{TR} = \underset{\sim}{u}_{R} - \underset{\sim}{u}_{FR} \tag{25}$$

where the appropriate radiation boundary conditions are given in terms of the x and y components as

$$\frac{\partial u^{x}_{TR}}{\partial x} = \frac{1}{c_s}\frac{\partial u^{x}_{TR}}{\partial t} \qquad \frac{\partial u^{y}_{TR}}{\partial y} = \frac{1}{c_s}\frac{\partial u^{y}_{TR}}{\partial t} \tag{26}$$

The resulting discretized equations of motion are then

$$\underset{\sim}{M}\,\ddot{\underset{\sim}{u}}_{R} + \underset{\sim}{K}\,\underset{\sim}{u}_{R} + [\,\underset{\sim}{P}_{S}\,(\dot{\underset{\sim}{u}}_{R} - \dot{\underset{\sim}{u}}_{FR})\,]_{\Gamma_S} + [\,\underset{\sim}{P}_{h}\,\dot{\underset{\sim}{u}}_{R}\,]_{\Gamma_h} = -\underset{\sim}{M}\,\ddot{\underset{\sim}{u}}_{IO} + [\,\underset{\sim}{S}\,]_{\Gamma_h}$$

$$\begin{aligned}\underset{\sim}{M}\,\ddot{\underset{\sim}{u}}_{R} + \underset{\sim}{K}\,\underset{\sim}{u}_{R} + [\,\underset{\sim}{P}_{S}\,\dot{\underset{\sim}{u}}_{R}\,]_{\Gamma_S} + [\,\underset{\sim}{P}_{h}\,\dot{\underset{\sim}{u}}_{R}\,]_{\Gamma_h} &= \\ -\underset{\sim}{M}\,\ddot{\underset{\sim}{u}}_{IO} + [\,\underset{\sim}{S}\,]_{\Gamma_h} + [\,\underset{\sim}{P}_{S}\,\dot{\underset{\sim}{u}}_{FR}\,]_{\Gamma_S}&\end{aligned} \tag{27}$$

where (for the case of vertically propagating seismic shear wave) the terms on the boundaries Γ_h and Γ_S are

$$\underset{\sim}{P}_{S} = \int_{\Gamma_S} \underset{\sim}{N}^{T} \begin{bmatrix} \frac{G}{c_S} & 0 \\ 0 & \frac{E}{c_P} \end{bmatrix} \underset{\sim}{N}\, d\Gamma \qquad \underset{\sim}{P}_{h} = \int_{\Gamma_h} \underset{\sim}{N}^{T} \begin{bmatrix} \frac{E}{c_P} & 0 \\ 0 & \frac{G}{c_S} \end{bmatrix} \underset{\sim}{N}\, d\Gamma$$

$$\underset{\sim}{S} = \int_{\Gamma_h} \underset{\sim}{N}^{T} \begin{bmatrix} 0 & 0 \\ 0 & \frac{G}{c_S} \end{bmatrix} \dot{u}_{IO}\, d\Gamma \tag{28}$$

It may be noted again that both u_R and u_{FR} are displacements relative to the incoming displacement at the truncation level.

Although a direct implementation of the above is possible two

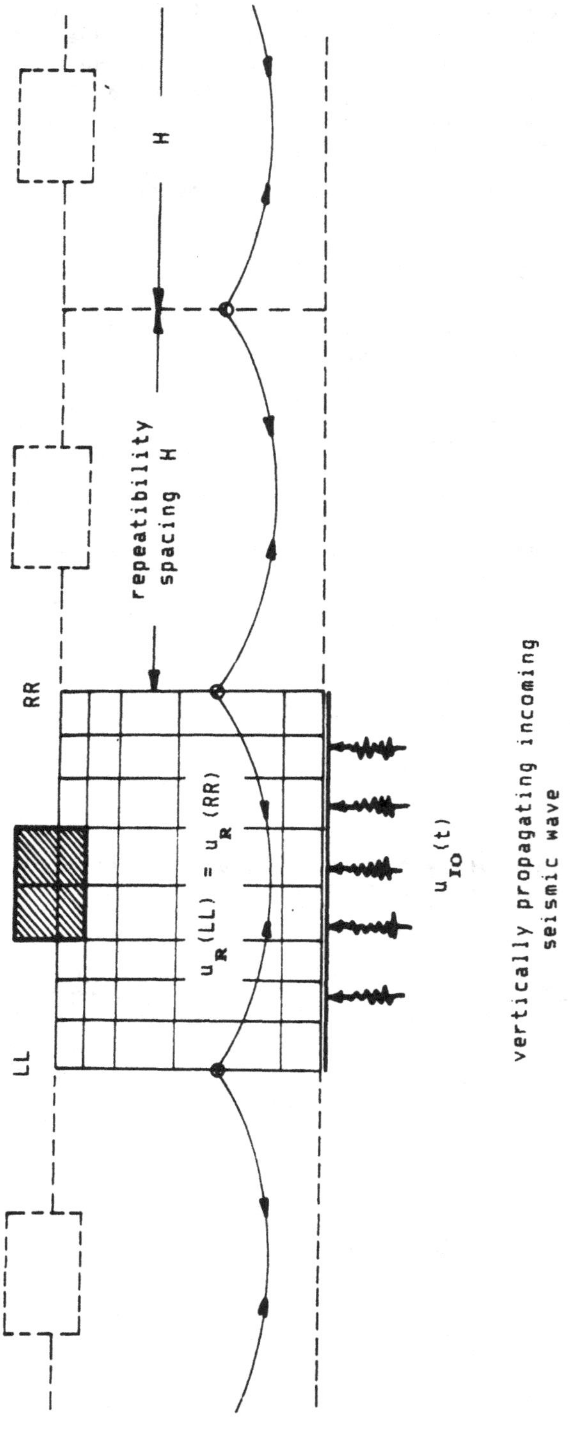

Fig 8 - "Tied" lateral boundaries approach

alternative procedures are suggested in the following sections and both are applicable to linear and nonlinear problems, providing the region near the truncation boundary remains elastic and homogeneous.

4.1. "Tied" lateral boundaries approach

The degrees of freedom for sections LL and RR are assumed to be "tied", implying the periodic nature of the surface perturbation, as if the same structure, fill or excavation is periodically repeated along the surface (Fig 8). It is evident from physical reasoning that correct results must be obtained if sufficiently large perturbation spacing H is adopted.

The problem can therefore be solved in an identical manner to the one dimensional case discussed previously, assuming that the incoming seismic signal does not vary spatially, i.e. it is the same signal for all nodes on the bottom truncation boundary.

The correctness of the procedure can easily be tested by solving simultaneously the one dimensional layered foundation problem ("free-field column") and comparing the results at sections LL and RR.

This approach is fully applicable to the nonlinear dynamic analysis and is preferred for its simplicity. The only domain where the linear material behaviour has to be maintained is the foundation layer immediately above the arbitrary bottom model truncation level.

4.2. Transmitting layer approach

The direct process implied in equations (28) and (29) can be represented by an alternative form, which is illustrated on Fig 9 - the finite element mesh is conceptually composed of three parts

(a) main mesh - degrees of freedom $\underset{\sim}{u}_R$

(b) free field column - degrees of freedom $\underset{\sim}{u}_{FR}$

(c) transmitting layer - degrees of freedom $\underset{\sim}{u}_{TR}$

From the problem definition it is clear that the motion of the foundation with the structure attached to it will differ from the free field motion. This difference in motion is due to the presence of the structure and can be viewed as the outward moving wave. It has already been said that the radiation (transmitting) boundary condition should here be imposed on the difference between the main mesh displacement and the free field displacement at the lateral truncation boundaries LL and RR.

This differential motion is clearly obtained as a difference between u_R and u_{FR}, as the incoming wave is the same forboth problems, i.e.

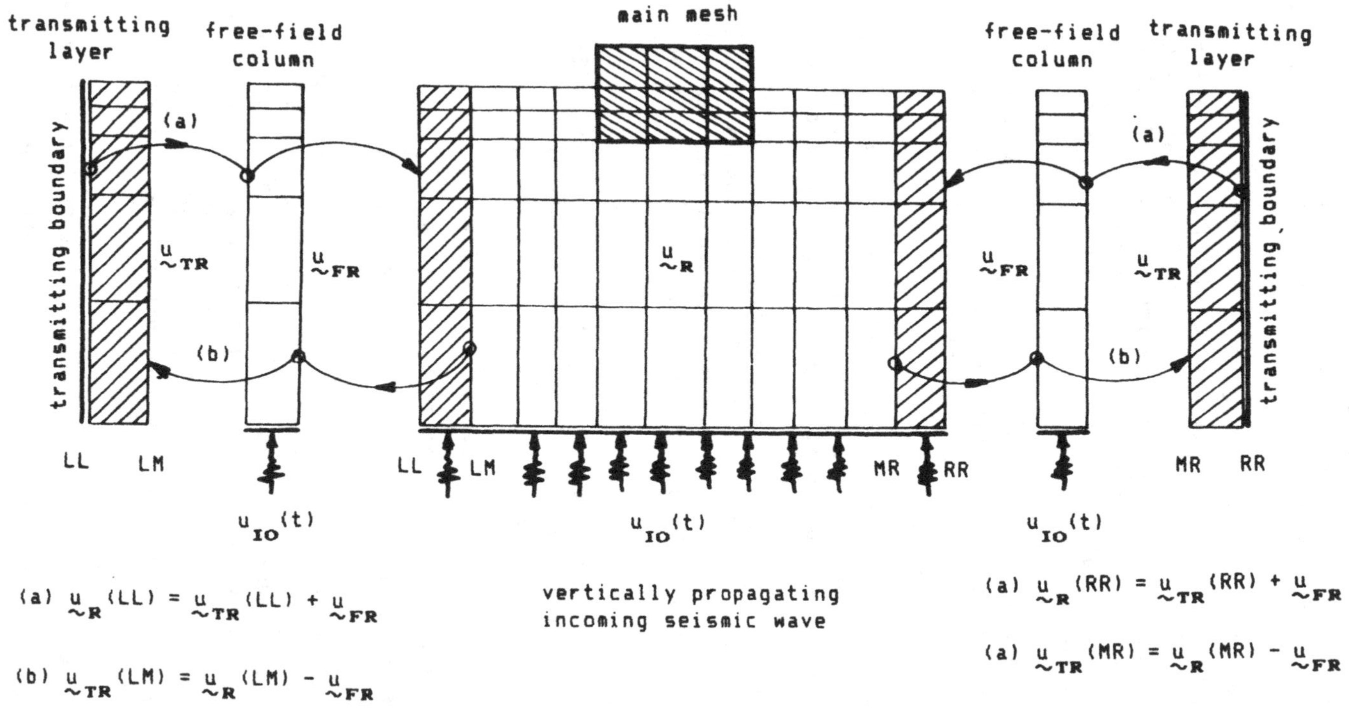

Fig 9 - Transmitting layer approach for the numerical treatment of lateral boundaries in earthquake problems

$$\underset{\sim}{u}_{TR} = \underset{\sim}{u}_{R} - \underset{\sim}{u}_{FR} \qquad (29)$$

The governing equations for the main mesh is given by

$$\underset{\sim}{M}^{M} \ddot{\underset{\sim}{u}}_{R} + \underset{\sim}{K}^{M} \underset{\sim}{u}_{R} = - \underset{\sim}{M}^{M} \ddot{\underset{\sim}{u}}_{IO} \qquad (30)$$

and the solution process follows standard algorithms.

The governing equations for the free-field column are similar to those of the main mesh and can be written as

$$\underset{\sim}{M}^{F} \ddot{\underset{\sim}{u}}_{FR} + \underset{\sim}{K}^{F} \underset{\sim}{u}_{FR} = - \underset{\sim}{M}^{F} \ddot{\underset{\sim}{u}}_{IO} \qquad (31)$$

The transmitting layer however, can be visualised as the last layer of elements along the lateral boundaries of the main mesh. These transmitting layers are used to transmit the perturbed motion caused by the presence of the structure. The appropriate governing equations for the transmitting layers are therefore homogeneous wave equations, which in the discretised form can be written as

$$\underset{\sim}{M}^{T} \ddot{\underset{\sim}{u}}_{TR} + \underset{\sim}{K}^{T} \underset{\sim}{u}_{TR} = 0 \qquad (32)$$

where u_{TR} is the differential motion defined in (29).

The actual computational procedure involves all three parts of the mesh simultaneoulsy. Within one computational step the following phases can be identified

(a) solve the main mesh system of equations (30), with u_R on the boundary LL and RR prescribed,

(b) solve the free-field column system of equations (31),

(c) solve the transmitting layer system of equations (32), with the difference between u_R and u_{FR} being applied at the boundary nodes ML and MR.

(d) obtain the new prescribed values for u_R on the boundary LL and RR of the main mesh by adding the free field column solution to the transmitting layer solution at boundary nodes LL and RR, and return to (a)

The above procedure is simplified in the actual computer implementation - the algorithm involves the solution of two problems in

parallel (main mesh and the free-field column) whereas the transmitting layer part amounts to merely special treatment of the lateral boundary nodes.
It should be noted again that all solution variables are displacements relative to the incoming vertically propagating seismic wave at the arbitrary truncation boundary located within the homogeneous bottom layer. The restriction of material linearity is here extended to the lateral transmitting layers as well as the soil layer above the bottom truncation boundary. In the model interior arbitrary non linear material behaviour can be accomodated.

4.3. Model problem in 2D - seismic excitation

Very simplified soil-structure model problem is illustrated in Fig 10 - homogeneous soil properties are assumed (with no bedrock) and various model truncations are considered for comparison purpose. Incoming seismic wave is assumed (El Centro) to propagate from infinity as a shear wave.

Three meshes considered (SN - shallow and narrow mesh, DN - deep and narrow mesh , SW - shallow and wide mesh) are analysed using both the "tied lateral boundaries" and "transmitting layer" approach. The comparison of displacement, acceleration and various stress histories of control points A and B of Fig 10 is given on Figs 11 to 13, and indicates the validity of the proposed formulation - the responses can hardly be distinguished from one another, and any differences can clearly be attributed to numerical dispersion.

Both "tied lateral boundaries" and "transmitting layer" procedures performed in the satisfatory way, with the clear attraction of the "tied lateral boundaries" approach being the simplicity of implementation.

5. CONCLUSIONS

It is believed that the direct treatment of earthquake input signal "entering" the model in the form of specified input waves is the logical way of dealing with the majority of foundation problems in earthquake engineering.

Numerous tests show the efficiency of the proposed procedure. The validity of the treatment of lateral boundaries by a repetitivity condition ("tied boundaries") is demonstrated, and as this procedure allows a full nonlinear behaviour it is suggested to employ this procedure as standard.

Similar, though more complex aproach can be adopted for the treatment of the inclined or horizontal incoming wave.

$$E = 2*10^9 \text{ Pa} \qquad \Delta t = 0.0075 \text{ sec}$$

$$\rho = 2*10^3 \text{ kg/m}^3 \qquad h = 10 \text{ m}$$

$$\nu = 0.30$$

possible
model truncations

vertically propagating incoming
seismic wave (El Centro)

Shallow and wide mesh   (SW)

Deep and narrow mesh   (DN)

Shallow and narrow mesh (SN)

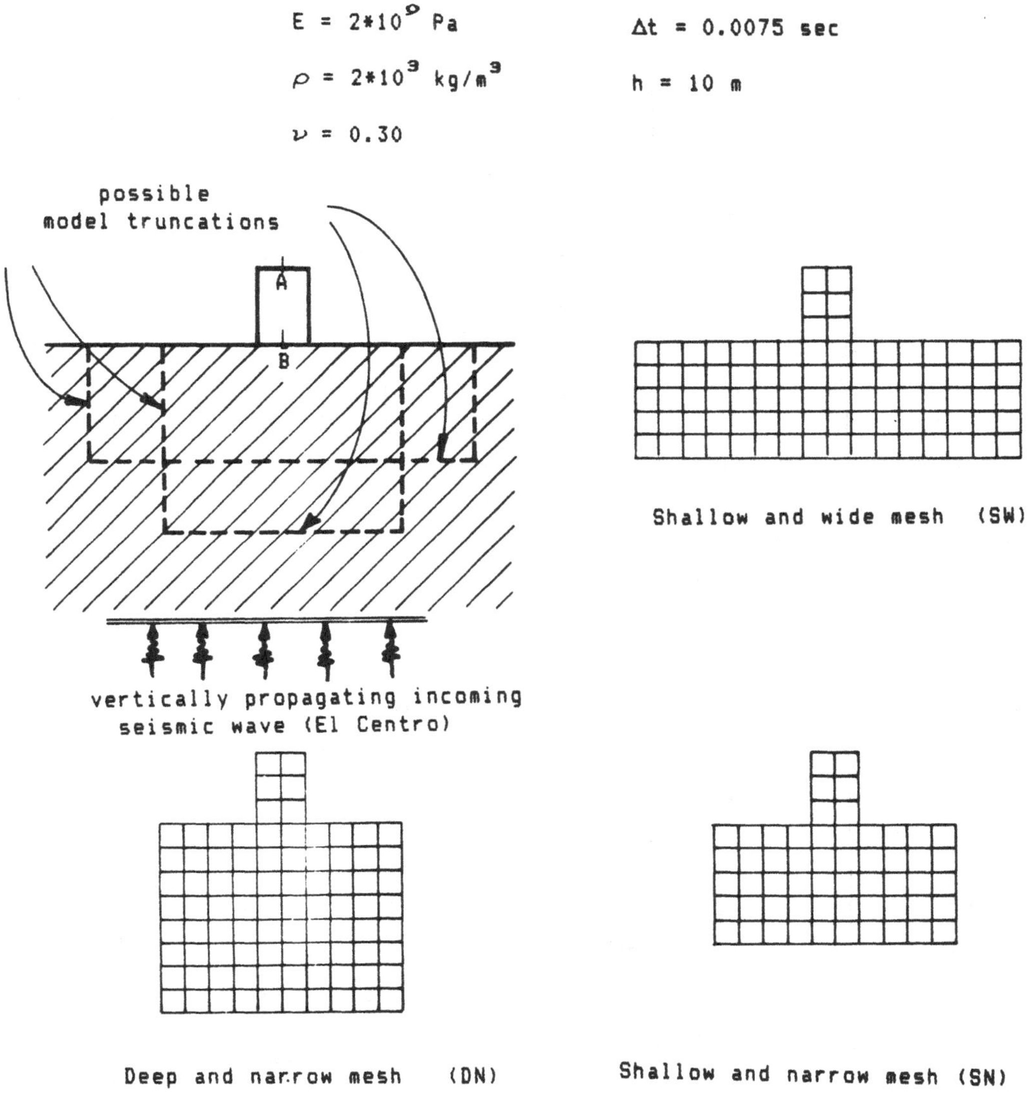

Fig 10 - Two dimensional model problem
and three meshes (SN, DN, SW)

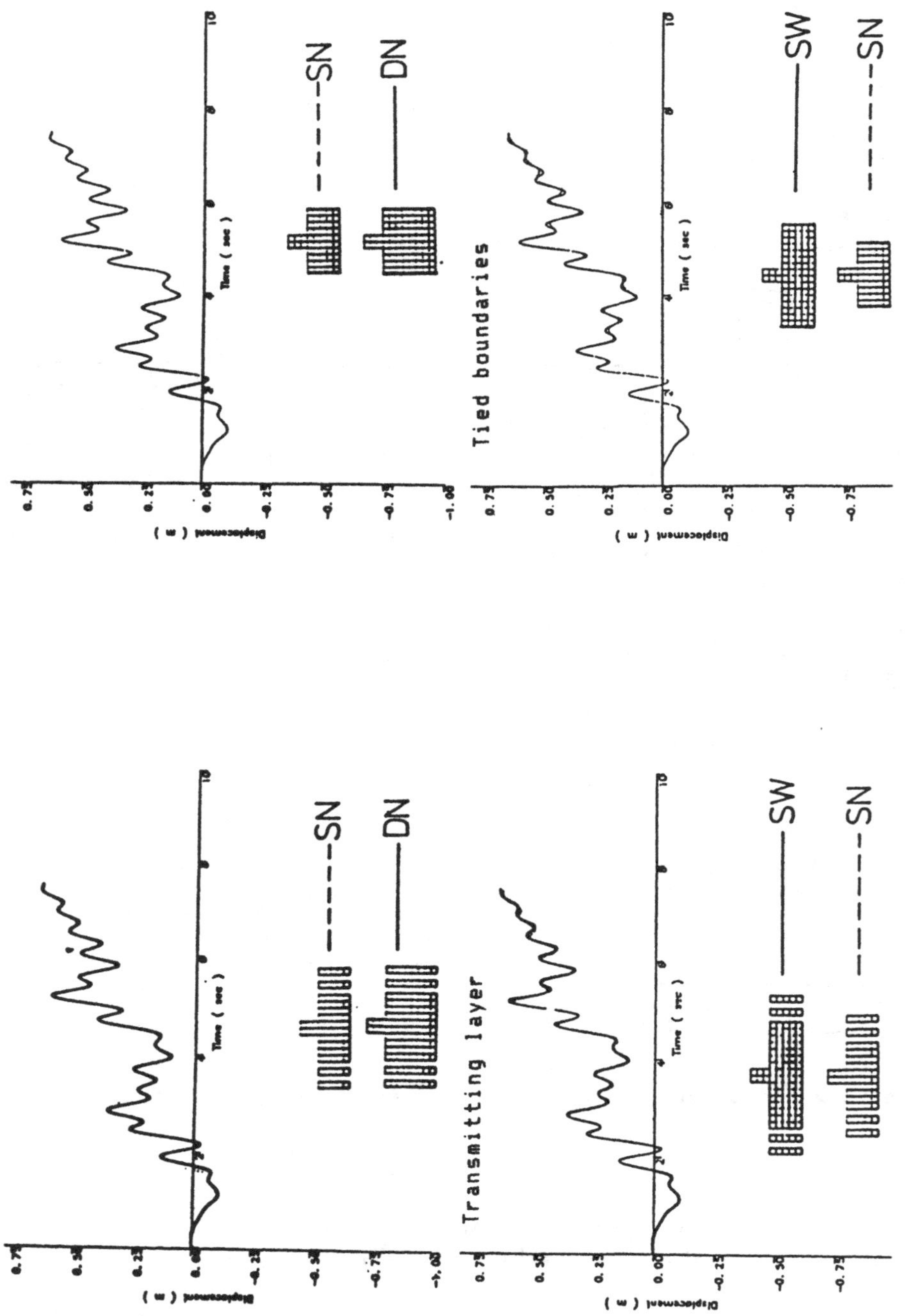

Fig 11a - Comparison of displacement histories (control point A) for various meshes

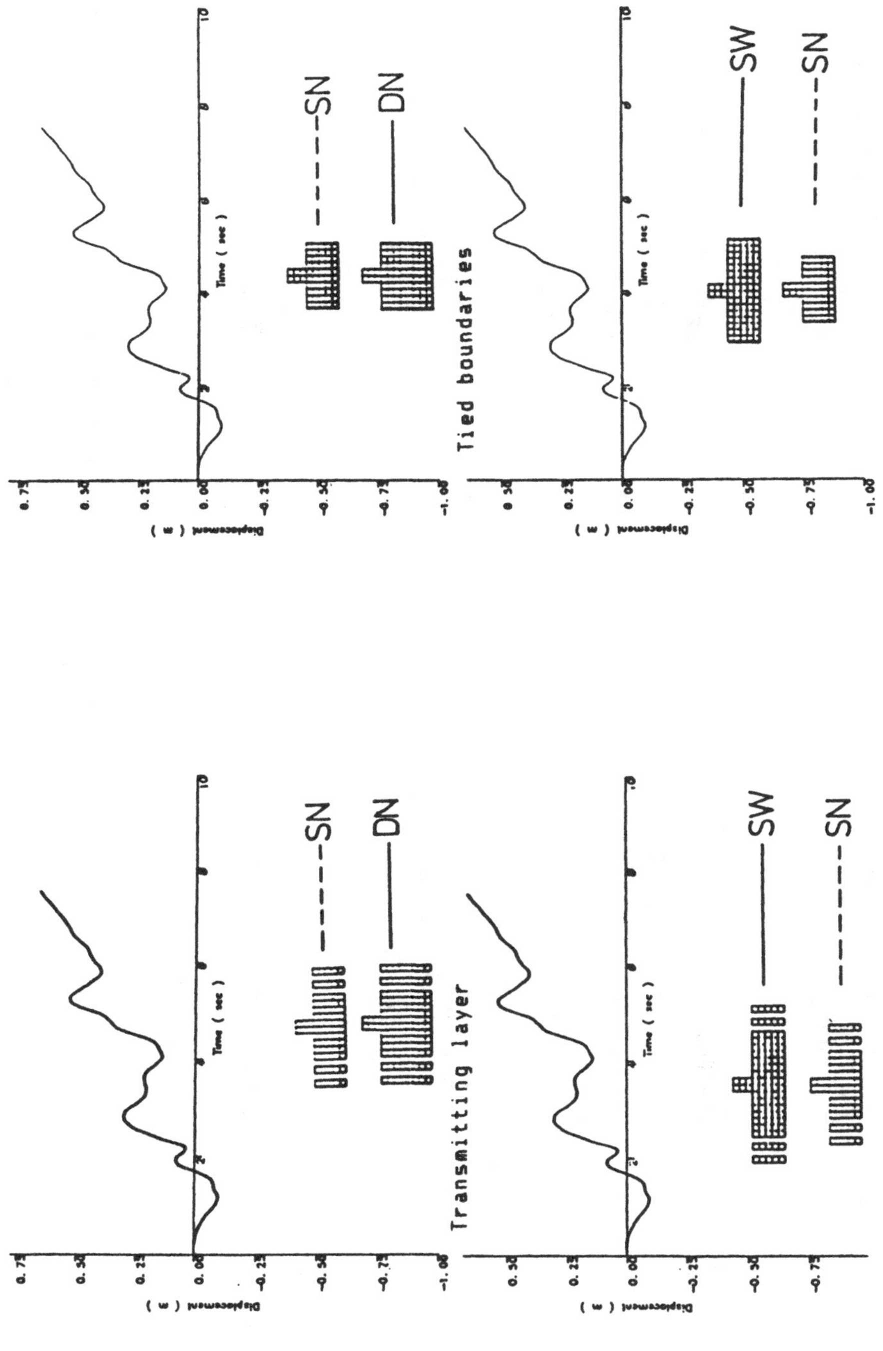

Fig 11b - Comparison of displacement histories (control point B) for various meshes

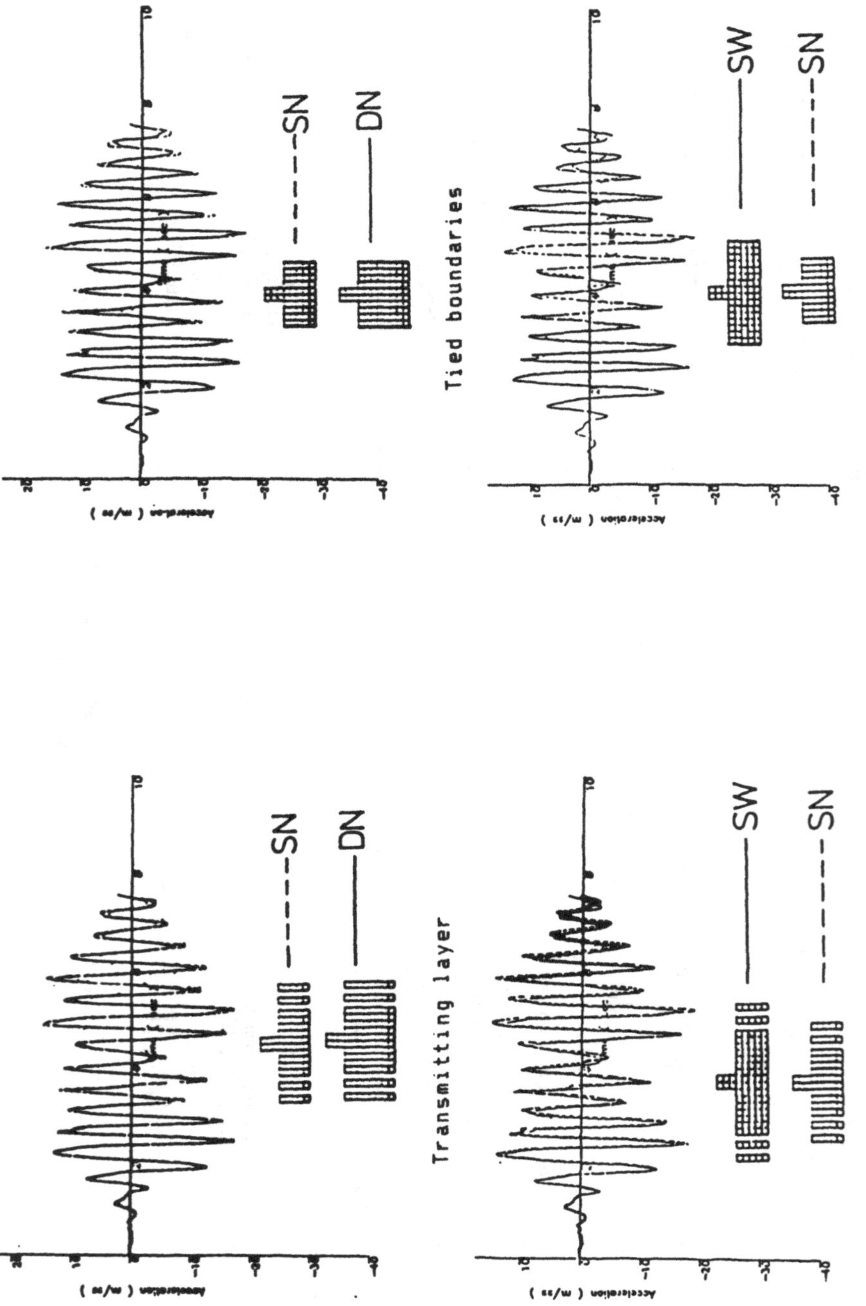

Fig 12a - Comparison of acceleration histories (control point A) for various meshes

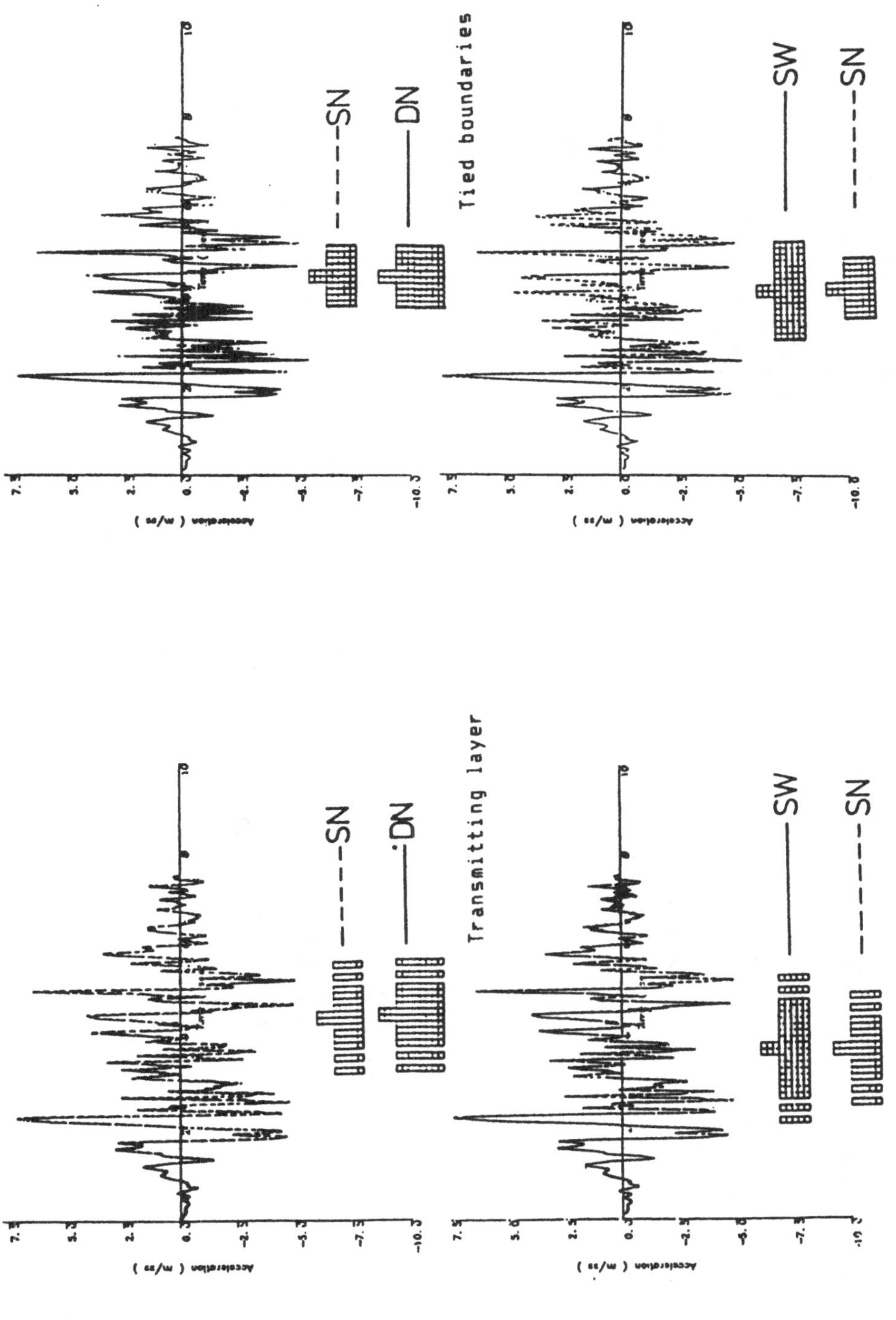

Fig 12b - Comparison of acceleration histories (control point B) for various meshes

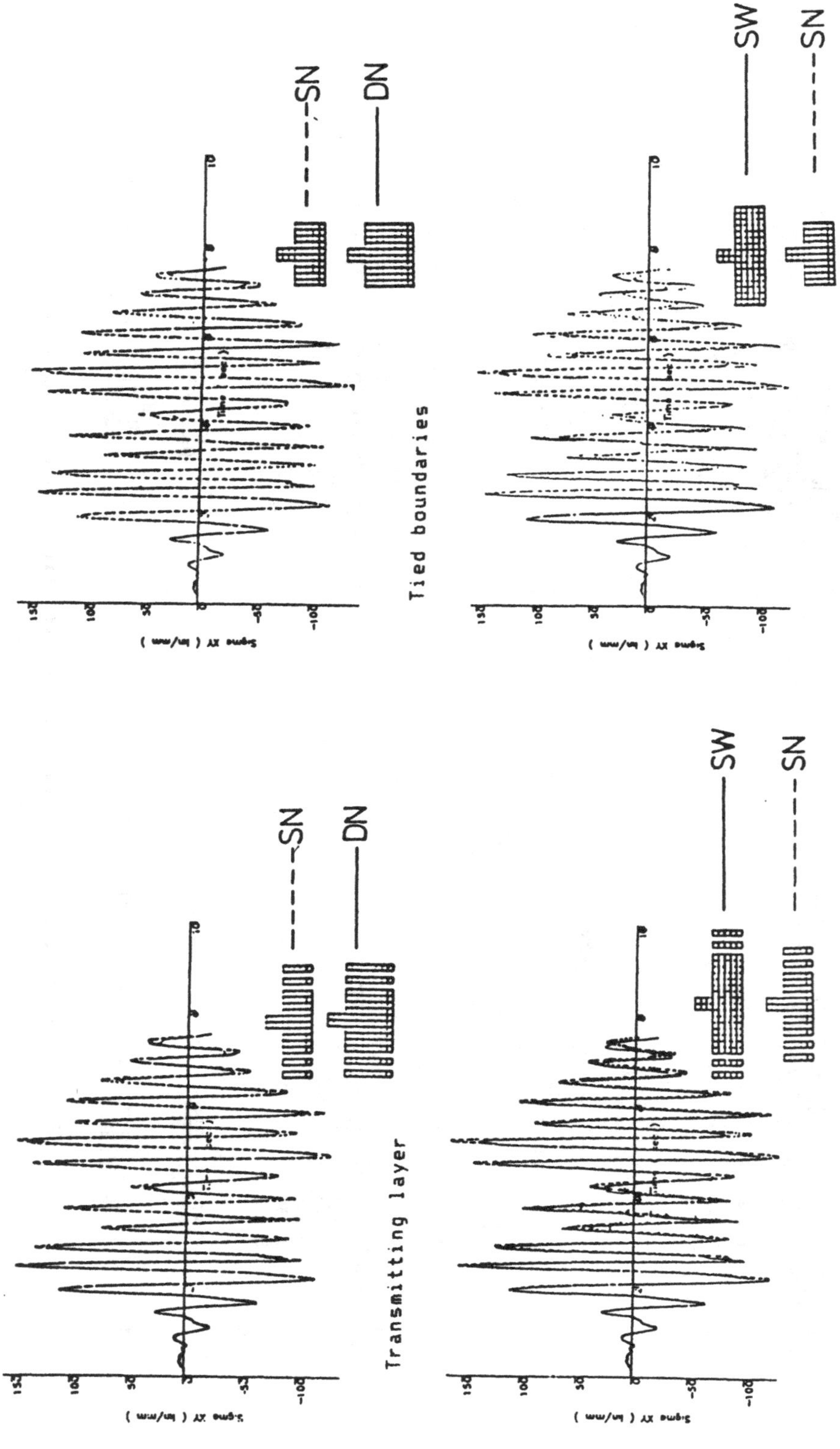

Fig 13a - Comparison of stress σ_{xy} histories (control point A) for various meshes

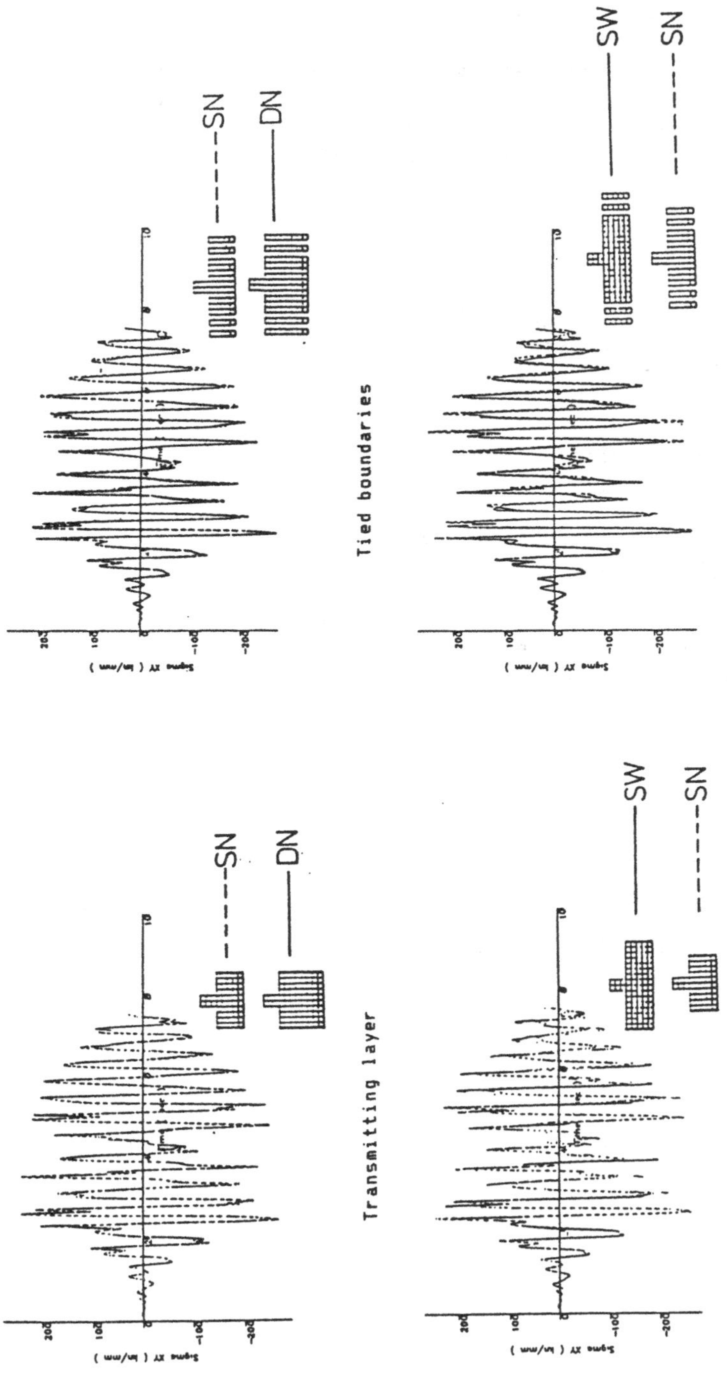

Fig 13b - Comparison of stress σ_{xy} histories (control point B) for various meshes

The procedure can equally well be applied to the problems where a distinct base rock or a soil layer with base rock like properties does exist - an appropriate (however small) proportion of the outgoing waves can be transmitted from the model. In the limit, when the bottom layer properties approach the infinitely rigid material, the conventional problem formulation is recovered.

6. REFERENCES

1. O.C.Zienkiewicz, R.W.Clough, H.B.Seed
   Earthquake analysis procedures for dams
   CIGB ICOLD Bulletin 52, 1986

2. O.C.Zienkiewicz
   The finite element method, Mc Graw Hill, 1977.

3. J.Lysmer, R.L.Kuhlmeyer
   Finite dynamic model for infinite media
   Journal ASCE, Eng. Mech. Div.
   Vol 95, No EM4 , August 1969, pp 859-877

4. W.White, S.Valiappan, I.K.Lee
   Unified boundary for finite dynamic models
   Journal ASCE, Eng. Mech. Div.
   Vol 103, No EM5 , Oct 1977, pp 949-965

5. W.Smith
   A non reflecting boundary for wave propagation problems
   Journal of Comp. Physics, 15 , 1973, pp 492-503

6. O.C.Zienkiewicz, N.Bicanic, F.Q.Shen
   Generalised Smith boundary - a transmitting boundary for
   dynamic computations,INME Swansea,C/R/570/1986

CHAPTER 4

COUPLED PROBLEMS AND THEIR NUMERICAL SOLUTION

O. C. Zienkiewicz, A.H.C. Chan
University College of Swansea, Swansea, UK

ABSTRACT:

As discussed in a previous publication [1] coupled problems in numerical analysis context can be divided into two main categories:

I) Those, where essentially different physics exist and in which the problem domain overlap completely or partially,

II) and those where coupling occurs only via an interface. Here the seperate domains may only differ either physically or only in the numerical technique used in each of them.

Some characteristics of the finite element discretization and steady state and transient solution technique are discussed and typical problems are illustrated. Special emphasis is given to some recent developments in transient dynamic problems including

1) Symmetrical direct solution techniques.
2) Simple Staggered solutions with unconditional stability.
3) Domain partition for single phase domain with Implicit-Implicit Staggered or Explict-Implicit Staggered Solution.

Stability analysis are also presented for the various solution techniques.

1. INTRODUCTION

Coupled problems appear frequently in the context of numerical (finite element) solution of physical problems. On occasion the physical description suffices to make clear the meaning of 'coupling', but frequently the definition is blurred. A previous publication has been devoted to an explicit definition of the term [1]. In this lecture, a brief reminder of the definition is given.

This states, the definition as given in reference [1]

'Coupled formulations are those applicable to multiple domains and dependent variables which usually (but not always) describible different physical phenomena and in which

a) neither domain can be solved correctly while seperated from the other,
b) neither set of dependent variables can be explicitly eliminated.'

An obvious 'coupled' problem is that of fluid-structure dynamic interaction. Clearly neither domain can be solved independently of the other and elimination of neither set of variables is possible at the differential equation level. On the other hand, such problems as steady state thermal stress phenomena are generally uncoupled in the above definition as the temperature distribution can be determined without the knowledge of the deformation and/or stress fields.

2. Classification of Coupled Problems

Two distinct categories of coupled problems exist, each amenable in general to similiar treatment:

Class I: Problems in which the various domain overlap (totally or partially only). Here the coupling occurs via the differential equations used to describe the different physical phenomena.

Class II: Problems in which the coupling occurs only at the domain interfaces and clearly this coupling will occur only through the interaction of the boundary condition imposed. Here two further subclassifications can be identified as

a) problems in which different physics (and/or dependent variables) occur in different domains,

b) problems in which identical physics and dependent variables are specified in different domains but differing numerical techniques are used in the individual domains.

The last category, (IIb), may indeed not correspond with the generally accepted concepts of 'coupled form', but it is convenient to include it here, representing, as it does, such procedures as substructing, some hydrid element forms, domain partitions, etc..

In many physical problems coupling may occur only in time-dependent (transient/dynamic) forms and disappear once the steady state conditions are reached. Once again a further subclassification into steady state (S) and transient (T) can be added.

At this stage, it is instructive to include some typical examples of various categories. To illustrate Class I, two typical problems are shown in Fig. 1.

The first one as illustrated in Fig. 1a is that of metal flow during extrusion process, where the yield conditions are strongly dependent on the temperature field. Here the equations in the domain of flow are influenced by the temperatures via viscosity and, in turn, the heat transfer equations have to include the heat generated by mechanical and convective transfer of work, thus coupling these effects to the flow equation [2,3]. The domains fo each of the physical equation overlap but clearly the heat transfer equation can include areas in which no flow occurs. Thus the overlapping is partial only. In this problem, both steady state (S) and transient (T) solutions are coupled.

The second example is one of considerable importance in civil engineering application, where a saturated soil mass is subject to dynamic excitation (e.g. that of earthquake motion [4,5]). Indeed coupling of three field can occur if the unsaturated zone is considered. Here the displacements and stresses of the soil skeletion are strongly affected by the pore pressure distribution and in turn the seepage phenomenon, giving the equations from which pore pressures can be found, is coupled to the rate at which displacements of the soil skeleton change and the fluid is squeezed in and out of the pores. Once again the two domains overloap partially - but where the steady state solution becomes completely decoupled. When such a steady state is reached, pressure field can be determined solely from the solution of the seepage equations.

Illustration of typical problems of Class II is shown in Fig. 2. Here, as a problem of class (IIa), we show one of interaction between a fluid and its structural container (fig. 2a). Clearly the two domains are seperate and interaction occurs only through the interface I. Furthermore, the problem is once again of category (T) as the static solution for the pressure can be found without reference to the structure.

A problem of class (IIb) category is shown in Fig. 2b. Here the different domains are all of a structural kind but which for some good reasons we may wish to treat them seperately. Such 'good reasons' may include

a) Simply the use of different finite element meshes in different domains (viz. Ω_1 and Ω_2)

b) The use of a different method of approximation in each of the domains (viz. Ω_3). In this illustration, for instance, it may be desirable to use a boundary solution technique so that Ω_3 can be extended to infinity.

c) In Transient (T) solution, it may be advisable to use different solution techniques such as Implicit and Explicit schemes in different regions of the domain.

3. The discretized forms of the Coupled problems

in this short survey it is inappropriate to deal with all the diverse coupled problems which have been successfully dealt with in the past. Nevertheless, a few examples are discussed in detail to illustrate some generally encountered features. Once again we shall follow the classification procedures previously discussed and discuss in the order of Class I, Class IIa and IIb. Though most realistic and useful solutions are nonlinear, only linear situations are discussed for simplicity sake. The nonlinear solutions shall only differ in the technique used for equation solving and procedures of such solution are described in literature.

3.1 Class I (soil pore fluid interaction)

soil - pore fluid interaction of the type illustrated in Fig. 1b is chosen here as example. The basic equations governing this problem are derived in reference [6] and can be written in indicial form (repeated dummy subscript implies summation unless otherwise stated).

$$\sigma_{ji,j}+\rho b_i-\rho \ddot{u}_i=0 \qquad (1)$$

(soil skeleton)

$$\varepsilon_{ij}=(u_{i,j}+u_{j,i})/2 \qquad (2)$$

$$d\sigma_{ij}=D_{ijkl}d\varepsilon_{kl}-\alpha\delta_{ij}dp \qquad \text{(constitutive relation)} \qquad (3)$$

$$\alpha=1-K_T/K_s \qquad (4)$$

$$\dot{p}/Q+\alpha\dot{\varepsilon}_{ii}+(k_{li}(-p_{,i}+\rho_f b_i-\rho_f \ddot{u}_i))_{,l}=0 \qquad (5)$$

$$1/Q=n/K_f+(\alpha-n)/K_s \qquad \text{(pore fluid)} \qquad (6)$$

σ_{ij} is the stress tensor

u_i is the displacement vector of the solid skeleton

ρ and ρ_f are the density of the mixture and fluid respectively

b_i is the body acceleration vector

ε_{ij} is the strain tensor

D_{ijkl} is the constitutive (fourth order) tensor

K_T, K_s and K_f are the Bulk modulus of the soil skeleton, soil grains and pore fluid respectively.

p is the pore pressure (compression positive)

k_{ij} is the permeability matrix

δ_{ij} is the Kronecker delta tensor

n is the porosity

Together with appropriate boundary condition in tractions, displacements, pressure and flow the system of eqn (1) to (6) specifies the problem completely. The reader can easily verify that the system is truly a coupled one in transient state as elimination will not result in a solvable set. With 'standard' Galerkin [7] type discretization, some interesting features can be observed.

Let the dependent variables be expressed by suitable shape functions and parameters (u,p) as

$$u_i=N\bar{u}_i \qquad \text{and} \qquad p=\tilde{N}\bar{p} \qquad (7)$$

then the standard Galerkin process will result in the following equation system assuming linearity:

$$M\ddot{\bar{u}}+K\bar{u}-Q\bar{p}=f_1 \qquad (8a)$$

$$S\dot{\bar{p}}+H\bar{p}+Q^T\dot{\bar{u}}=f_2 \qquad (8b)$$

Here M and K are the usual forms of mass and stiffness matrices occuring in an uncoupled solid elastic problem (corresponding to the behaviour of the soil skeleton) and S and H are the compressibility and conductivity matrices corresponding to the uncoupled seepage problem. The exact form of these matrices is well known (viz. reference [7]) and need not be repeated here - but the fact that they are symmetric positive definite matrices is important. The only matrix which generates the coupling is the Q matrix.

Clearly it is advisable to use the same finite element domain for each of the variables but, equally clearly, it is not necessary to use the same order of approximation or the same number of nodes. Indeed, same order of interpolation may introduce spurious oscillations when undrained condition is approached. It can be shown that in this particular problem higher effeciency and good accuracy usually will result from the use of lower order interpolation for p and one order higher for u.

The discretized equation system (8) is a transient second order system which in combined form can be written as

$$\begin{bmatrix} M & 0 \\ 0 & 0 \end{bmatrix}\begin{bmatrix} \ddot{\bar{u}} \\ \ddot{\bar{p}} \end{bmatrix} + \begin{bmatrix} 0 & 0 \\ Q^T & S \end{bmatrix}\begin{bmatrix} \dot{\bar{u}} \\ \dot{\bar{p}} \end{bmatrix} + \begin{bmatrix} K & -Q \\ 0 & H \end{bmatrix}\begin{bmatrix} \bar{u} \\ \bar{p} \end{bmatrix} = \begin{bmatrix} f_1 \\ f_2 \end{bmatrix} \qquad (9)$$

We note immediately the asymmetry of the coefficient matrices and if standard time stepping procedure is used for the combined system difficulties will be encountered.

Although the formulation given is a particular one, two general points applicable to many Class I problems will be noted.

1) The discretization presents no difficulties and follow precisely the pattern used for mixed problems.

2) The combined equation system frequently presents asymmetries.

3.2 Class II(a)

The typical example here is the one of fluid-structure interaction illustrated in Fig. 2a.

The full specification of such coupled equations can be found elsewhere [8,9], but it is obvious that the structural formulation and discretization will involve as the basic variable the displacement u_i, while that of the fluid (assumed to be frictionless, i.e. accoustic) will have as the basic variable the pressure p.

On the interface I the coupling is specified by requiring that

a) the traction on the solid is

$$t_i=-n_i p \tag{10}$$

where n_i is the outward normal unit vector, and

b) the outward normal gradient of p for the fluid field is

$$\bar{n}_i p_{,i}=-\rho n_j u_j \tag{11}$$

where $\bar{n}_i$ is the outward normal of the fluid domain. The first boundary condition is well known to be a natural one for a displacement discretization of the structural domain. Again, using the standard (Galerkin) procedure and discretizing

$$u_i=N\bar{u}_i \quad \text{and} \quad p=\tilde{N}\bar{p} \tag{12}$$

we can write

$$M\ddot{\bar{u}}+K\bar{u}-Q\bar{p}=f_1 \tag{13}$$

Q is defined for the interface variables only, so the matrix is extremely sparse.

In the fluid domain the differential equation governing the pressure distribution is the Helmholz one (c being the velocity of sound),

$$p_{,ii}=\ddot{p}/c^2 \tag{14}$$

for which condition (11) is again a natural one. Standard discretization will lead to

$$S\ddot{\bar{p}}+H\bar{p}+Q^T\ddot{u}=f_2 \tag{15}$$

We observe that the structure of equation (13) and (15) is similiar (but not identical) to those of the previous example. Furthermore, if the two equations are written as a second-degree system, we note similiar dissymmetries to those occuring in (9)

$$\begin{bmatrix} M & 0 \\ Q^T & S \end{bmatrix} \begin{bmatrix} \ddot{\bar{u}} \\ \ddot{\bar{p}} \end{bmatrix} + \begin{bmatrix} K & -Q \\ 0 & H \end{bmatrix} \begin{bmatrix} \bar{u} \\ \bar{p} \end{bmatrix} = \begin{bmatrix} f_1 \\ f_2 \end{bmatrix} \qquad (16)$$

The equations in the above undamped form poses real eigenivalues which however, cannot be easily computed by a standard algorithm. Indeed methods of computations for such eigenvalues are at least twice as expensive as those for standard symmetric positive definite systems [10] and much effort has been put in towards recasting of the coupled problem in a more amenable form [11,12]. However, such a symmetrization can only be accomplished at the cost of introducing additional variables - and its cost effectiveness is by no means obvious.

Summarizing we note that in a Class II(a) problem it appears once again that the discretization and coupling have been produced in a 'natural' manner with no particular difficulties and once again similiar equations to those arising in Class I problems are obtained.

The main difference lies in the fact that the coupling terms involve only an interface integral in which both interpolation functions N and $\tilde{N}$ occur. It is obviously convenient to divide the interface into similiar elements but indeed different approximation orders i.e. different number of nodes can be used. It is for instance convenient in problems where the influence of the fluid is of secondary importance to use a coarser discretization for the fluid then for the structure shown in Fig. 3. Now the 'interface' term is simply evaluated as an 'element' corresponding to the fluid nodes 5 and 6 coupling with the structural nodes 1, 2, 3 and 4.

It will be observed that the penalty for the use of the coarser mesh in the fluid domain is the increased bandwidth of the coupling - a feature of no importance if staggered solution processes described in the next section are used.

3.3 Class II b

The final type of coupling problem is, in which identical variables (but not necessary the discretization or the timestepping procedures) occur in seperate domain. The discussion for domain subdivision has been dealt with in detail in reference [1]. In this section, the heat conduction problem is used as an example to illustrate the possible time domain solution partitions. The governing equation [7] can be expressed as

$$(k_{ij}T_{,j})_{,i}+Q-\rho c\dot{T}=0 \tag{17}$$

where k_{ij} is the conductivity matrix,
T is the temperature field,
ρ is the density of the material,
Q is the internal source function,
c is the specific heat capacity.

As can be noted that similiar governing equation can result for fluid diffusion problem. Using standard discretization procedure:

$$T=N\bar{T} \tag{18}$$

The result is a first degree equation system.

$$S\dot{\bar{T}}+H\bar{T}=f \tag{19}$$

where S and H are the capacity and conductivity matrix respectively. Both of them are symmetric positive definite and diagonally dominant matrices. This can, in fact be casted into the simpliest coupled equation. Letting

$$\bar{T}=[T_1,T_2]^T \quad (20a) \qquad f=[f_1,f_2]^T \quad (20b)$$

$$S=\begin{bmatrix} S_{11} & S_{12} \\ S_{21} & S_{22} \end{bmatrix} \quad (20c) \qquad H=\begin{bmatrix} H_{11} & H_{12} \\ H_{21} & H_{22} \end{bmatrix} \quad (20d)$$

Then eqn (19) can be rewritten into its coupled form

$$\begin{bmatrix} S_{11} & S_{12} \\ S_{21} & S_{22} \end{bmatrix}\begin{bmatrix} \dot{T}_1 \\ \dot{T}_2 \end{bmatrix}+\begin{bmatrix} H_{11} & H_{12} \\ H_{21} & H_{22} \end{bmatrix}\begin{bmatrix} T_1 \\ T_2 \end{bmatrix}=\begin{bmatrix} f_1 \\ f_2 \end{bmatrix} \tag{21}$$

Note, in particular that unlike the Class I and IIa problem the equation is still symmetric in its coupled form as $S_{12}=S_{21}^T$ and $H_{12}=H_{21}^T$.

Using the SSpj [13] time stepping schemes, we can discretize in time with

$$T_1=T_1^{n-1}+\alpha_1^n t' \quad (22a)$$

$$T_2=T_2^{n-1}+\alpha_2^n t' \quad (22b)$$

where $t'=t-t_{n-1}$ and $t_{n-1}\leq t\leq t_n$. Using different weighting functions for the two equations, we have

$$\begin{bmatrix} S_{11}+H_{11}\Theta\Delta t & S_{12}+H_{12}\Theta\Delta t \\ S_{21}+H_{21}\bar{\Theta}\Delta t & S_{22}+H_{22}\bar{\Theta}\Delta t \end{bmatrix} \begin{bmatrix} \alpha_1^n \\ \alpha_2^n \end{bmatrix} = \begin{bmatrix} F_1^n \\ F_2^n \end{bmatrix} \quad (23)$$

If $\Theta=\bar{\Theta}$, then we will recover the ordinary explicit ($\Theta=0$) or Implicit ($\Theta\neq0$) form. For $\Theta\geq1/2$, the scheme is unconditionally stable. For $\Theta<1/2$, the critical time step is given by

$$\Delta t_{crit}\leq 2/(\omega_{max}(1-2\Theta)) \quad (24)$$

where ω_{max} is the maximum eigenvalue of the generalized eigenvalue problem

$$HX=\omega CX \quad (25)$$

with $\Theta\neq\bar{\Theta}$, various interesting possibilities can be obtained.

3.3.1 Implicit-Implicit Direct Solution

With $\Theta>1/2$, the SSpj scheme exhibits numerical damping, which is useful in the suppression of spurious numerical oscillations. However, with $\Theta=1/2$, the SSpj first order scheme is of $O(\Delta t^2)$ accuracy (i.e. error in the solution is of the order Δt^2), so it may be desirable to use different Θ's in different part of the domain. In order to assess the stability of such partition, the error of the previous solution must not be allowed to grow away from the exact solution. The following stability analysis procedure can be established for the Linearized equation. (For nonlinear equation, the subtraction of error has to be done using the Taylor expansion)

a) constant terms like f_1 in eqn (21) are neglected as they are unchanged even if the solution of last step is incorrect.

b) the current solution is expressed as a multiple of the previous solution i.e.

$$T_1^n = \lambda T_1^{n-1} \tag{26}$$

where λ is a complex multiplicative parameter. The scheme is stable if $|\lambda|>1$ and if $|\lambda|=1$, it must not be a multiple root for oscillation may occur about this root [14]. Applying the well-known conformal mapping in the complex plane

$$\lambda=(1+z)/(1-z) \tag{27}$$

which maps the unit disc to the negative real plane. Then $|\lambda|\lesssim 1$ imples $Re(z)\lesssim 0$ for which the famous Routh-Hurwitz criteria can be used [15]. As

$$T_1^n = T_1^{n-1} + \alpha_1^n \Delta t \tag{28a}$$

so

$$\alpha_1^n = 2zT_1^{n-1}/((1-z)\Delta t) \tag{28b}$$

eqn (23) now becomes

$$\begin{bmatrix} (S_{11}+H_{11}\theta\Delta t)2z/(1-z)+H_{11}\Delta t & (S_{12}+H_{12}\theta\Delta t)2z/(1-z)+H_{12}\Delta t \\ (S_{21}+H_{21}\bar{\theta}\Delta t)2z/(1-z)+H_{21}\Delta t & (S_{22}+H_{22}\bar{\theta}\Delta t)2z/(1-z)+H_{22}\Delta t \end{bmatrix} \cdot \begin{bmatrix} T_1^{n-1} \\ T_2^{n-1} \end{bmatrix} = \begin{bmatrix} 0 \\ 0 \end{bmatrix} \tag{29}$$

It is convenient at this stage to consider an equivalent scalar problem, to simplify the consideration to a single eleemnt. For non-trivial solution of eqn (29), it is necessary that the determinant be identically zero

$$\begin{vmatrix} (s_{11}+h_{11}\theta\Delta t)2z+h_{11}\Delta t(1-z) & (s_{12}+h_{12}\theta\Delta t)2z+h_{12}\Delta t(1-z) \\ (s_{21}+h_{21}\bar{\theta}\Delta t)2z+h_{21}\Delta t(1-z) & (s_{22}+h_{22}\bar{\theta}\Delta t)2z+h_{22}\Delta t(1-z) \end{vmatrix} = 0 \tag{30}$$

and now the symmetry property of S and H implies that $s_{12}=s_{21}$ and $h_{12}=h_{21}$. Expanding the determinant in eqn (30), we arrived at the second order characteristic equation or the Routh-Hurwitz polynomial.

$$f(z)=a_0z^2+a_1z+a_2=0 \tag{31a}$$

$$a_0=(2s_{11}+h_{11}\Delta t(2\theta-1)).(2s_{22}+h_{22}\Delta t(2\bar{\theta}-1)) - (2s_{12}+h_{12}\Delta t(2\theta-1)).(2s_{21}+h_{21}\Delta t(2\bar{\theta}-1)) \quad (31b)$$

$$a_1= (2s_{11}+h_{11}\Delta t(2\theta-1)).h_{22} + (2s_{22}+h_{22}\Delta t(2\bar{\theta}-1)).h_{11} -(2s_{12}+h_{12}\Delta t(2\theta-1)).h_{21} - (2s_{21}+h_{21}\Delta t(2\bar{\theta}-1)).h_{12} \quad (31c)$$

$$a_2= h_{11}h_{22}-h_{21}h_{12} \quad (31d)$$

The Routh-Hurwitz criterion for $Re(z)\leq0$ requires that $a_0>0$ and a_1, $a_2>0$. In Galerkin procedure, the resulting matrix is always diagonal dominaced symmetric positive definite i.e. s_{11}, $s_{22}>|s_{12}|=|s_{21}|$ and so does matrix H. Now all three coefficients of the polynomial are positive so the Routh-Hurwitz criterion is satisfied when $\theta\geq1/2$ and $\bar{\theta}\geq1/2$. The scheme is therefore unconditionally stable.

3.3.2 Explicit-Implicit (Staggered) Partition

Using simple lumping procedure [7] via nodal integration or otherwise, S_{11} can be made diagonal and S_{12}, S_{21} are identically zero. To eliminate the contribution of H_{12} and H_{21} in the first set of equation in eqn (23), θ can be chosen to be zero.

Explicit-Implicit partition scheme is particularly useful if a part of the structure (domain) is especially softer than the remaining portion. An efficient algorithm can be produced by using an explicit scheme for the 'softer' region and implicit scheme for the stiffer part. With this kind of partitioning, computational effort can be reduced without excessive restriction on the critical time step. Some examples of application has been presented in reference [16].

With $\theta=0$, eqn (23) now becomes

$$\begin{bmatrix} S_{11} & 0 \\ H_{21}\bar{\theta}\Delta t & (S_{22}+H_{22}\bar{\theta}\Delta t) \end{bmatrix} \begin{bmatrix} \alpha_1^n \\ \alpha_2^n \end{bmatrix} = \begin{bmatrix} F_1^n \\ F_2^n \end{bmatrix} \quad (32)$$

As no contribution of α_2^n is found in the first set of equations, α_1^n can be determined first then α_2^n can be determined from the second equation. Thus the equation can be solved in a naturally 'staggered' manner, with the two equations set evaluated one after the other, less effort is involved in the solution of the combined set. Since the two equations are not solved at the same time, the solution process seems to move from one set of equations to another (i.e. in a staggering manner).

The Routh-Hurwitz polynomial for this partition is again second order with

$$a_0 = (2s_{11}-h_{11}\Delta t).(2s_{22}+h_{22}\Delta t(2\bar{\theta}-1))+h_{12}\Delta t.h_{21}\Delta t(2\bar{\theta}-1) \quad (33a)$$

$$a_1 = (2s_{11}-h_{11}\Delta t).h_{22}+(2s_{22}+h_{22}\Delta t(2\bar{\theta}-1)).h_{11}+h_{12}\Delta t h_{21} \quad (33b)$$
$$-h_{21}\Delta t(2\bar{\theta}-1)h_{12}$$

$$a_2 = h_{11}h_{22}-h_{12}h_{21} \quad (33c)$$

The criteria for stability is

$$a_0, a_1 > 0 \text{ if } 2s_{11}-h_{11}\Delta t > 0 \text{ and } \bar{\theta} \geq 1/2 \quad (34)$$

and a_2 is always positive. Therefore, the time step is restricted by

$$\Delta t_{crit} \leq 2s_{11}/h_{11} \text{ or } 2/\omega_{max} \quad (35)$$

where ω_{max} is determined from the generalized eigenvalue problem

$$H_{11}X = \omega S_{11}X \quad (36)$$

which can be considerably bigger than that of total explicit scheme if the partition is properly chosen as outlined in the above section.

3.3.3 Implicit-Implicit Staggered Scheme

A similiar staggered scheme can be developed for the Implicit-Implicit method also. Rewriting equation (23) in its staggered form, and the unknown T_2^n is extrapolated from the previous known value T_2^{n-1}, so that the first set of equations can be solved without the knowledge of the second

$$(S_{11}+H_{11}\theta\Delta t)a_1^n = F_1^n \quad (37a)$$

$$(S_{22}+H_{22}\bar{\theta}\Delta t)a_2^n = F_2^n-(S_{21}+H_{21}\bar{\theta}\Delta t)a_1^n \quad (37b)$$

The stability polynomial of the scalar equation is once again of second order

$$f(z) = a_0z^2+a_1z+a_2 \quad (38a)$$

$$a_0 = (2s_{11}+h_{11}(2\theta-1)\Delta t).(2s_{22}+h_{22}(2\bar{\theta}-1)\Delta t)$$
$$+h_{12}\Delta t.(2s_{21}+h_{21}(2\bar{\theta}-1)\Delta t) \qquad (38b)$$

$$a_1 = (2s_{11}+h_{11}\Delta t(2\theta-1)).h_{22}+(2s_{22}+h_{22}\Delta t(2\bar{\theta}-1)).h_{11}$$
$$-(2s_{21}+h_{21}\Delta t(2\bar{\theta}-1)).h_{12}+h_{21}\Delta t.h_{12}\Delta t \qquad (38c)$$

$$a_2 = h_{11}h_{22}-h_{12}h_{21} \qquad (38d)$$

The Routh-Hurwitz Criterion is satisfied if θ and $\bar{\theta}$ are greater than 1/2. The scheme is more effecient thatn the Implicit-Implicit Direct Solution but now numerical damping is intrinsically present.

4 Numerical Solution of Discretized coupled equations

Once the discretization (or semi-discretization in space only) of a coupled system has been achieved, two main possibilities exist for its numerical solution. In the first a direct solution of the combined system is carried out and in the second the solution is staggered with each component treated as an uncoupled set and the coupling variables suitably extrapolated.

4.1 Symmetriced direct solution for time stepping procedure

In this process, symmetrization of an asymmetric system such as that of eqn (9) and (16) is accompanied by a rearrangement of equations after the temporal approximation. Detail of the derivations are given by Zienkiewicz and Taylor [17].

Advantages of this method are:

1. Unlike in standard time stepping procedures, a suitable multiplying constant is applied to one of the sets of equation, thus resulting in symmetric matrices in the solution.

2. The solution process is unconditionally stable without modifying the constituent matrices.

3. Same order of interpolation is not required when the variables contain different order of the derivatives in time. For example, in eqn (9), the order of time derivatives for u_i is second order and p is only of first order.

4.1.1 Class I problem

Using the soil-pore fluid problem as an example again, we now use lowest possible approximation in time with the SSpj time stepping scheme for u_i and p.

$$u=u_n+\dot{u}_n t'+\alpha_n t'^2/2 \qquad (39a)$$

$$p=p_n+\beta_n t' \qquad (39b)$$

(on noting that coefficient for the second derivative of p terms are zero). The weighted equation becomes

$$M.\alpha_n+K.(u_n+\dot{u}_n\theta_1\Delta t+\alpha_n\theta_2\Delta t^2/2)-Q.(p_n+\beta_n\theta_1\Delta t)=f_1 \qquad (40a)$$

$$Q^T.(\dot{u}_n+\alpha_n\bar{\theta}_1\Delta t)+S\beta_n+H.(p_n+\beta_n\bar{\theta}_1\Delta t)=f_2 \qquad (40b)$$

or

$$\begin{bmatrix} M+K\theta_2\Delta t^2/2 & -Q\theta_1\Delta t \\ Q^T\bar{\theta}_1\Delta t & S+H\bar{\theta}_1\Delta t \end{bmatrix} \begin{bmatrix} \alpha_n \\ \beta_n \end{bmatrix} = \begin{bmatrix} F_1 \\ F_2 \end{bmatrix} \qquad (41)$$

The equation can be symmetriced by multiplication of a simple constant $-\theta_1/\bar{\theta}_1$ to the second set of equations. Indeed the same procedure can be applied to GNpj [18] scheme (the Generalized Newmark or Beta-m method). The scheme is unconditionally stable if

$$\theta_2 \geq \theta_1 \geq 1/2 \quad \text{and} \quad \bar{\theta}_1 \geq 1/2 \qquad (42)$$

Detail of derivation of the condition are presented in the appendix of reference [17]. It can be shown that such symmetrization is possible no matter what order of (permissible) interpolation is used for either variable.

4.1.2 Class IIa problem

Here we consider a typical structure fluid interaction problem given by eqn (16). The second order SSpj is similiarly applied to both u_i and p in the equation with

$$u=u_n+\dot{u}_n t'+\alpha_n t'^2/2 \qquad (43a)$$

$$p=p_n+\dot{p}_n t'+\beta_n t'^2/2 \qquad (43b)$$

going through similiar procedure as in previous section, the final equation is

$$\begin{bmatrix} M+K\theta_2\Delta t^2/2 & -Q\theta_2\Delta t^2/2 \\ Q^T & S+H\bar{\theta}_2\Delta t^2/2 \end{bmatrix} \begin{bmatrix} \alpha_n \\ \beta_n \end{bmatrix} = \begin{bmatrix} F_1 \\ F_2 \end{bmatrix} \qquad (44)$$

The equation can be symmetriced by the mulitplier $-\theta_2\Delta t^2/2$. Once again, the stability condition is typical that of implicit solution

$$\theta_2 \geq \theta_1 \geq 1/2 \text{ and } \bar{\theta}_2 \geq \bar{\theta}_1 \geq 1/2 \qquad (45)$$

The detail of the derivation can also be found in the previous reference [17]. Once again, the symmetrization process is applicable to the GNpj scheme and all order of interpolation (except first order one as it is not permissible).

4.2 Modified Staggered procedure

In the 'staggered' solution procedure widely discussed by Park and Felippa [19,20,21] each of the equation of the combined system is treated seperately to determine a single field variable set. With coupling being accomplished in the first approximation by expolating the second equation variable is determined directly by its solution.

We illustrate this with the soil-pore fluid example. We would first compute displacements by solving a dynamic single phase system of u_i, using an extrapolation of p. Next we use the computed u_i to solve the second (seepage) equation for p.

Such staggered is however only conditionally stable even if unconditionally stable algorithms are used in each set and hence a modification to the system need to be introduced. Now it is possible that

1. With minor modifications of the constituent matrices, the modified staggerd procedure is unconditionally stable whereas the ordinary staggered procedure is not [22].

2. Many well established programs exist for the solution of uncoupled processes such as occur in parts of the coupled system (viz. eqn (9), and (16)). Such programs can be used directly in a staggered process if a slight modifications of entering the coupling terms as a forcing function and some matrices are introduced. This can avoid the necessity of writing complex, new program for every new coupled situation encountered.

3. Iteration of nonlinearity can be limited to one field and this can reduce the cost considerably. Nevertheless, accuracy is improved if all the fields are involved in the iterations.

4. In most practical problems a considerable degree of nonlinearity will be encounterd and some iteration is generally necessary. The iterative structure of treating uncoupled problems seperately allows for this non-linearity admirably.

5. The staggered solution involves matrices with smaller bandwidth. When the coupling is slight, the efficiency is much better than the direct method.

Ordinary Implicit-Implicit staggered solution has been considered by Paul [16]. Explicit-Implicit staggered solution for soil-pore fluid equation has been considered by Leung [23] and Zienkiewicz et al [24].

4.2.1 Unconditionally stable staggered solution procedure for soil-pore fluid interaction problems [25]

Various schemes which lead to an unconditionally stable solution of eqn (9) has been proposed in reference [25] with numerical examples. Only the Newmark-Newmark scheme is derived below as an example.

If the permeability approaches zero (i.e. the undrained condition is reached) then the contribution of permeability matrix H vanishes and in that case the pressure can be obtained by integrating the second equation in eqn (9). (noting that in such condition $f_2=0$)

$$p_{n+1}-p_n=-S^{-1}Q^T(u_{n+1}-u_n) \tag{46}$$

The equivalent 'force' due to this additional pressure on the soil skeleton would be from eqn (8b)

$$Qp_{n+1}=-QS^{-1}Q^T(u_{n+1}-u_n)+p_n \tag{47}$$

Now, the equations (8a) are modified to consider this increased pressure, a staggered scheme of the form given below appears natural.

$$Mu_{n+1}+(K_d+QS^{-1}Q^T)u_{n+1}=f_{n+1}+Qp_n+QS^{-1}Q^Tu_n \quad (48)$$

or in general, u_n and p_n can be replaced by p^p_{n+1} and u^p_{n+1} with superscript p denoting predicted values. The next equation needs no modification as u_{n+1} can be obtained from eqn (8a). Indeed this staggered solution with suitably predicted values of p^p_{n+1} and u^p_{n+1} together with suitable integration formulae turns out to be unconditionally stable when stable ranges of those formulae are used for each component equation. The selection of formulae and predictors for stable solution is outlined in the later part of this section.

Although the process appears as simple mathematical manipulation, it has a physical significance as the new stiffness matrix is simply the 'undrained' stiffness matrix K_u, defined as:

$$K_u=K_d+QS^{-1}Q^T \quad (49)$$

In this analysis therefore, the displacements are obtained by assuming the undrained material behaviour rather than drained behaviour and the corresponding pressure rise simply dissipates by diffusion.

with C_o - continuity for p-discretization, the computation of $QS^{-1}Q^T$ is expensive and time consuming but an alternative approximation for K_u can be obtained by using piecewise continuous p dicretization and reduce integration for the penalty type procedure [7]. The evaluation of K_u by this means is no more complex than K_d. Now the evaluation of $QS^{-1}Q^T$ is not made and the approximation results in

$$K_u=K_d+QS^{-1}Q^T \simeq \int B^T(D_d+Qmm^T)B\,d\Omega \quad (50)$$

where m is the vector equivalent of the kronecker delta, D_d is the drained modulus matrix, B is the strain displacement matrix.

If such a computational process is adopted, it is necessary to evaluate the last term of eqn (48) in a consistent manner (i.e. using the same approximation for $QS^{-1}Q^T \simeq K_u-K_d$). The procedure improves the performance of the approximation for nearly undrained behaviour.

It is of importance to note that similiar approaches have been intuitively used by some soil mechanicians for computing fluid behaviour as undrained for relatively long period of several time steps and then applying drainage dissipation.

The stability of a partitioned scheme depends on the following factors:

1. criteria of partitioning of the field
2. the integration formula used
3. the selected predictors and
4. the computational path selected.

In the problem, the partitioning is obtained according to the physics of the problem. The computational path is determined by the approximated procedures outlined above. Though an alternative path exists, we shall restrict our discussion to this computational path only. For some integration formula, the alternative computational is the only path that unconditionally stability can be achieved. As a typical example, the SS22 for displacement and SS11 for pressure is used as the integration formula. The predicter formula is then fixed by the previous choices and the unconditionally stable requirement.

After applying the SSpj scheme to eqn (48) and (8b), the resulting equation is

$$M\alpha_n+(K_d+QS^{-1}Q^T)u_{n+\alpha}=f_1+QS^{-1}Q^Tu^p_{n+\alpha}+Qp^p_{n+\alpha} \quad (51a)$$

$$Q^T\bar{\dot{u}}_{n+\alpha}+S\beta_n+H\bar{p}_{n+\alpha}=\bar{f}_2 \quad (51b)$$

where $u_{n+\alpha}=u_n+\dot{u}_n\theta_1\Delta t+\alpha_n\theta_2\Delta t^2/2$ (51c)

$$u^p_{n+\alpha}=u_n \quad (51d)$$

$$\bar{\dot{u}}_{n+\alpha}=\dot{u}_n+\alpha_n\bar{\theta}_1\Delta t \quad (51e)$$

$$p^p_{n+\alpha}=p_n \quad (51f)$$

$$\bar{p}_{n+\alpha}=p_n+\beta_n\bar{\theta}_1\Delta t \quad (51g)$$

Applying the standard stability analysis procedure to the scalar equivalent of equation set (51). The Routh-Hurwitz polynomial is of third order:

$$f(z)=a_0z^3+a_1z^2+a_2z+a_3 \quad (52)$$

for $Re(z) \leq 0$ the requirement is

$$a_0 > 0, \quad a_1, a_2, a_3 \geq 0 \tag{53}$$

$$\begin{vmatrix} a_1 & a_3 \\ a_0 & a_2 \end{vmatrix} > 0$$

In this case

$$a_0 = (4m + k_d \Delta t^2 (2\theta_2 - 2\theta_1) + q^2 \Delta t^2 / s (2\theta_2 - 2\theta_1)).(2s + h\Delta t (2\bar{\theta}_1 - 1)) - 2q^2 \Delta t^2 (2\bar{\theta}_1 - 1) \tag{54a}$$

$a_0 > 0$ if $m, k_d, h, s > 0$, $\bar{\theta}_1 \geq 1/2$, $\theta_2 \geq \theta_1$ and $2(\theta_2 - \theta_1) \geq 2\bar{\theta}_1 - 1$

$$a_1 = (4m + k_d \Delta t^2 (2\theta_2 - 2\theta_1) + q^2 \Delta t^2 / s (2\theta_2 - 2\theta_1)).h\Delta t + (k_d \Delta t^2 (2\theta_1 - 1) + q^2 \Delta t^2 / s (2\theta_1)).(2s + h\Delta t (2\bar{\theta}_1 - 1)) + 2q^2 \Delta t^2 (2\bar{\theta}_1 - 1) - 2q^2 \Delta t^2 \tag{54b}$$

$a_1 > 0$ if $\theta_1 \geq 1/2$ together with the previous conditions

$$a_2 = (k_d \Delta t^2 (2\theta_1 - 1) + q^2 \Delta t^2 / s (2\theta_1)).h\Delta t + k_d \Delta t^2.(2s + h\Delta t (2\bar{\theta}_1 - 1)) + 2q^2 \Delta t^2 \tag{54c}$$

$a_2 > 0$ if previous conditions are fulfilled

$a_3 = k_d \Delta t^2.h\Delta t$ is always greater than zero (54d)

For the determinant condition $a_1 a_2 - a_0 a_3 > 0$, it is fulfilled for $\theta_2 = \theta_1 = \bar{\theta}_1 = 1/2$ and the condition with general coefficients is checked with numerical procedure, no violation has been observed yet. This scheme has been used with great success by Zienkiewicz et al [25 ,27] and Paul and Zienkiewicz [26]. Alternative stabilized staggered procedure has been proposed by Park [22] but is involves significantly more computational efforts.

4.2.2 Unconditionally stable staggered solution procedure for fluid-structure interaction problem

Similiar method of modification can be applied to the fluid-structure interaction problem as outlined in eqn (16). First the eqn (16) is rewritten as

$$M\ddot{u}+K u+QS^{-1}Q^{T}u=f_1+QS^{-1}Q^{T}u^{p}+Qp^{p} \tag{55a}$$

$$Q^{T}\dot{u}+S\dot{p}+Hp=f_2 \tag{55b}$$

Applying the SS22 scheme to both variables:

$$M\alpha_n+(K+QS^{-1}Q^{T})u_{n+\alpha}=f_1+QS^{-1}Q^{T}u^{p}_{n+\alpha}+Qp^{p}_{n+\alpha} \tag{56a}$$

$$Q^{T}\alpha_n+S\beta_n+H\bar{p}_{n+\alpha}=\bar{f}_2 \tag{56b}$$

where $u_{n+\alpha}=u_n+\dot{u}_n\theta_1\Delta t+\alpha_n\theta_2\Delta t^2/2$ (56c)

$$u^{p}_{n+\alpha}=u_n+\dot{u}_n\theta_1\Delta t \tag{56d}$$

$$p^{p}_{n+\alpha}=p_n+\dot{p}_n\theta_1\Delta t \tag{56e}$$

$$\bar{p}_{n+\alpha}=p_n+\dot{p}_n\bar{\theta}_1\Delta t+\beta_n\bar{\theta}_2\Delta t^2/2 \tag{56f}$$

The scalar equivalent of equation set will lead to a fourth order Routh-Hurwitz polynomial for stability analysis

$$f(z)=a_0z^4+a_1z^3+a_2z^2+a_3z+a_4 \tag{57}$$

The condition for stability is

$$a_0>0,\ a_1,a_2,a_3,a_4\geq 0 \tag{58a}$$

$$\begin{vmatrix} a_1 & a_3 \\ a_0 & a_2 \end{vmatrix}>0, \qquad \begin{vmatrix} a_1 & a_3 & 0 \\ a_0 & a_2 & a_4 \\ 0 & a_1 & a_3 \end{vmatrix}>0 \tag{58b}$$

With

$$a_0=(4m+(2\theta_2-2\theta_1)k_d\Delta t^2+2\theta_2q^2\Delta t^2/s).(4s+h\Delta t^2(2\bar{\theta}_2-2\bar{\theta}_1)) \\ -8\theta_1q^2\Delta t^2>0 \tag{59a}$$

$a_0 > 0$ if $m, k_d, h, s > 0$ and $\theta_2 \geq \theta_1, \bar{\theta}_2 \geq \bar{\theta}_1$

$$a_1 = (4m + (2\theta_2 - 2\theta_1)k_d\Delta t^2 + 2\theta_2 q^2\Delta t^2/s).h\Delta t^2(2\bar{\theta}_1 - 1)$$

$$+(2\theta_1 - 1)k_d\Delta t^2.(4s + h\Delta t^2(2\bar{\theta}_2 - 2\bar{\theta}_1)) + 4(2\theta_1 - 1)q^2\Delta t^2 \quad (59b)$$

$a_1 \geq 0$ if $\theta_1 \geq 1/2$, $\bar{\theta}_1 \geq 1/2$ with previous conditions

$$a_2 = (4m + (2\theta_2 - 2\theta_1)k_d\Delta t^2 + 2\theta_2 q^2\Delta t^2/s).h\Delta t^2 \quad (59c)$$

$$+(2\theta_1 - 1)k_d\Delta t^2.h\Delta t^2(2\bar{\theta}_1 - 1) + k_d\Delta t^2.(4s + h\Delta t^2(2\bar{\theta}_2 - 2\bar{\theta}_1)) + 4q^2\Delta t^2$$

$a_2 \geq 0$ if all the previous conditions are satisfied

$$a_3 = (2\theta_1 - 1)k_d\Delta t^2.h\Delta t^2 + k_d\Delta t^2.h\Delta t^2(2\bar{\theta}_1 - 1) \quad (59d)$$

$a_3 \geq 0$ is okay

$a_4 = h\Delta t^2.k_d\Delta t^2$ is always greater than zero (59e)

When $\theta_2 = \theta_1 = \bar{\theta}_2 = \bar{\theta}_1 = 1/2$, $a_1 = a_3 = 0$, the condition (58b) degenerates to

$$a_2^2 - 4a_0a_4 > 0 \quad (60)$$

and this is fulfilled. Numerical procedure is used to assess the condition with general coefficients and no violation has been found. $K+QS^{-1}Q^T$ has no significant physical meaning in this case and no simplification as the soil-pore fluid case is available. However, it is noted that the matrix $QS^{-1}Q^T$ contains non-zero terms only for the elements on the interface and is extremely sparse.This matrix can be obtained by the method of influence and the cost is much smaller than full iversion of the matrix.

5. Concluding Remarks

In the preceding sections, we have outlined some of the recent developments in the solution of coupled problem. The research group in the Institute has been engaged in the solution h of numerous 'coupled' problems nad a few of these will be listed here, and examples shown to demonstrate the usefulness of the methods.

Fig. 4 represents a problem of electro-osmotic seepage in which the pore water flow equations are influenced by an applied electrostatic potential [28]. The problem is linear and of Class I (S).

Fig. 5 gives an example of thermally coupled metal flow which occurs in metal rolling. The obvious practical interest in such problems stresses the importance of effecient solutions. A high degree of non-linearity is presented in such problems due to non-Newtonian behaviour of the fluid as well as due to the complex coupling coefficients.

Fig. 6 indicates that thermal flow coupling also occurs in thermally induced convection. This figure, taken from reference [29], shows a stream function plot for flow induced purely by a temperature difference applied to the boundary. Both problems are of Class I (S).

Fig. 7 to 9 illustrates an important application of coupled soil-pore fluid interaction during an earthquake shock applied to a dam-reservoir system in a centrifuge experiment performed in Cambridge [30]. The layout of the experiment and the finite element idealization is shown in Fig. 7. The nonlinear soil response causes a pore pressure build up and sustantial deformation has occured at the crest of the dyke. This is a problem of Class I (T). Fig. 8 (taken from reference [31]) represents a solution by the symmetrized SSpj u-p formulation for the problem. Fig. 9 (taken from reference [25]) illustrated the same problem calculated using the modified staggered scheme which is unconditionally stable. Both compares well with each other and the experimental results.

The last example in Fig. 10 shows how coupling of finite element and boundary solutions can be applied to a practical situation. We illustrate this by an example involving a magnetic field where the exterior region is modelled by a boundary solution for ease of dealing with an infinite domain [32]. This is a classical (IIb) problem.

The few examples quoted just give some ideas of the variety of problems which involve coupling. Many other situations such as electromechanical problems, geothermal energy, etc., have been dealt with extensively by others. The field of activity is here so large that it is impossible to delve into details. More new developments are still under consideration by our group. We hope, however, that the present paper will give some basic guidelines and suggestions useful to those working in some coupling condition.

6. ACKNOWLEDGEMENT

The second author (A.H.C. Chan) would like to express his thanks to the Croucher Foundation of Hong Kong for their finanical support which made his study in Swansea possible.

7. REFERENCES

1. Zienkiewicz, O.C.: Coupled Problems and their Numerical Solution, Chapter 1 in: Numerical Methods in Coupled Systems (Eds. R.W. Lewis, P. Bettess and E. Hinton), John Wiley and Sons Ltd, 1984, pp. 35-58.
2. Zienkiewicz, O.C., Onate, E. and Heinrich, J.C.: Plastic flow in metal forming. 1. Coupled thermal behaviour in extrusion. 2. Thin sheet forming. in: Proceedings of the Winter Annual Meeting of the American Society of Mechanical Engineers on 'Applications of Numerical Methods to Forming Processes', San Francisco, Dec., 1978, ASME, New York, 1979, pp. 107-120.
3. Zienkiewicz, O.C., Onate, E. and Heinrich, J.C.: A general formulation for coupled thermal flow of metals using finite elements., Int. J. Num. Meth. Engrg., 17 (1981), 1497-1514.
4. Zienkiewicz, O.C., Chang, C.T. and Bettess, P.: Drained, undrained, consolidating and dynamic behaviour assumptions in soil. Limits of validity. Geotechnique, 30 (1980), 385-395.
5. Zienkiewicz, O.C., Chang, C.T., Hinton, E. and Leung, K.H.: Earth dam analysis for earthquakes. Numerical solution and constitutive relations for nonlinear (damage) analysis. Paper presented to the Conference on Design of Dams to Resist Earthquakes, Institution of Civil Engineers, London, 1-2 October 1980.
6. Zienkiewicz, O.C.: The coupled problems of soil-pore fluid-external fluid interaction: Basis for a general geomechanics code. in Proceedings of the Fifth International Conference on Numerical Methods in Geomechanics, Nagoya, 1-5 Apr., 1985, 1731-1740.
7. Zienkiewicz, O.C.:The Finite Element Method, third Edn., McGraw-Hill, New York, 1977.
8. Zienkiewicz, O.C. and Bettess, P.: Fluid-structure dynamic interaction and wave forces. An introduction to numerical treatment. Int. J. Num. Meth. Engrg., 13 (1978), 1-16.
9. Zienkiewicz, O.C. and Newton, R.E.: Coupled vibrations of a structure submerged in a compressible fluid. Paper presented to the International Symposium on Finite Element Techniques, Stuttgart, 1-15 May 1969.
10. Gupta, K.K.: On a numerical solution of the supersonic panel flutter eigenproblem. Int. J. Num. Meth. Engrg., 10 (1976), 637-645.
11. Ohayon, R. and Valid, R.: True symmetric formulation of free vibrations of fluid-structure interaction - applications and extensions. In Proceedings of the Conference on Numerical Methods for Coupled Problems (Eds. E. Hinton et al), Pineridge Press, Swansea, 1981, 335-345.
12. Ohayon, R.: Symmetric variational formuations for harmonic vibration problems by coupling primal and dual variables - applications to fluid-structure coupled systems. La Recherche Aerospatiale, 3 (1979), 69-77.

13. Zienkiewicz, O.C., Wood, W.L., Hine, N.W. and Taylor, R.L.: A unified set of sinle step algorithms Part I: General formuation and applications, Int. J. Num. Meth. Engrg., 20 (1984), 1529-1552.
14. Lambert, J.D.: Computational Methods in Ordinary Differential Equations. John Wiley and Sons Ltd., 1973.
15. Gantmacher, F.R.: Applications of the theoy of matrices, Second part of the 'A theory of matrices', translated and revised by J.L. Brenner et al, Interscience Publishers, Inc., 1959.
16. Paul, D.K.: Efficient Dynamic Solutions for single and coupled multiple field problems. Ph.D. Thesis C/Ph/64/82, University College of Swansea, 1982.
17. Zienkiewicz, O.C. and Taylor, R.L.: Coupled Problems - A simple time-stepping procedure. Comm. Appl. Num. Meth., 1 (1985), 233-239.
18. Katona, M.G. and Zienkiewicz, O.C.: A unified set of single step algorithms Part 3: The Beta-m method, a generalization of the Newmark scheme. Int. J. Num. Meth. Engrg., 21 (1985), 1345-1359.
19. Felippa, C.A. and Park, K.C.: Staggered transient analysis procedures for coupled mechanics systems: formulation. Comp. Meth. Appl. Mech. Engrg., 24 (1980), 61-111.
20. Park, K.C.: Partitioned transient analysis procedures for coupled field problems: stability analysis. J. Appl. Mech., 47 (1980), 370-376.
21. Park, K.C. and Felippa, C.A.: Partitioned transient analysis procedures for coupled field problems: accuracy analysis. J. Appl. Mech., 47 (1980), 919-926.
22. Park, K.C.: Stablization of partitioned solution procedure for pore fluid-soil interaction analysis. Int. J. Num. Meth. Engrg., 19 (1983), 1669-1673.
23. Leung, K.H.: Earthquake response of saturated soils and liquefaction. Ph.D. Thesis C/Ph/79/84, University College of Swansea, 1984.
24. Zienkiewicz, O.C., Hinton, E., Leung, K.H. and Taylor, R.L.: Staggered, Time Marching Schemes in Dynamic Soil Analysis and Selective Explicit Extrapolation Algorithms. in Procedings of Conference on Innovative Numerical Analysis for the Engineering Sciences (Eds. R. Shaw et al), University of Virginia Press, 1980.
25. Zienkiewicz, O.C., Paul, D.K. and Chan, A.H.C.: Unconditionally stable staggered solution procedure for soil-pore fluid interaction problems. Research Report, C/R/586/87, Aug., 1987, University College of Swansea, submitted to Comm. Appl. Num. Meth..
26. Paul, D.K. and Zienkiewicz, O.C.: Earthquake and post earthquake response validity of liquefiable soils using centrifuge test. Research report, C/R/585/87, Aug., 1987, University College of Swansea.

27. Zienkiewicz, O.C., Paul, D.K. and Chan, A.H.C.: Numerical solution of saturated flat sand bed response when subjected to earthquake using centrifuge model test. Research Report, C/R/587/87, Aug., 1987, University College of Swansea.
28. Lewis, R.W. and Gerner, R.W.: A finite element solution of coupled electrokinetic and hydrodynamic flow in porous media. Int. J. Num. Meth. Engrg., 5 (1972), 41-56.
29. Marshall, R.S., Heinrich, J.C. and Zienkiewicz, O.C.: Natural convection in a square enclosure by a finite element penalty function method using primitive fluid variables. J. Num. Meth. Heat Transfer, 1 (1978), 315-330.
30. Venter, K.V.: KVV03 data report: Revised data report of a centrifuge model test and two triaxial tests. Andrew N. Schofield and Associate, Oct., 1985.
31. Chan, A.H.C.: A unified Finite Element Solution to Static and Dynamic problems of Geomechanics. Ph.D. Thesis, University College of Swansea, in preparation, 1988.
32. Zienkiewicz, O.C., Kelly, D.W. and Bettess, P.: Marriage a la mode - the best of both worlds (finite elements and boundary integrals. Paper presented to the International Symposium on Innovative Numerical Analysis in Applied Engineering Science, Versailles, May 1977. Also In Energy Methods in Finite Element Analysis (Eds. R. Glowinski, E.Y. Rodin and O.C. Zienkiewicz), John Wiley and Sons Ltd., Chichester, 1979, 81-106.

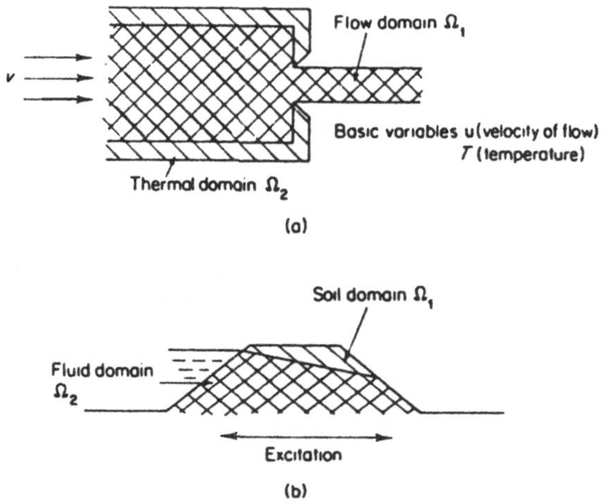

Fig. 1 Typical Class I problems: (a) Class I (S) or (T) metal extrusion temperature-dependent properties; (b) Class I (T) soil-pore fluid interaction.

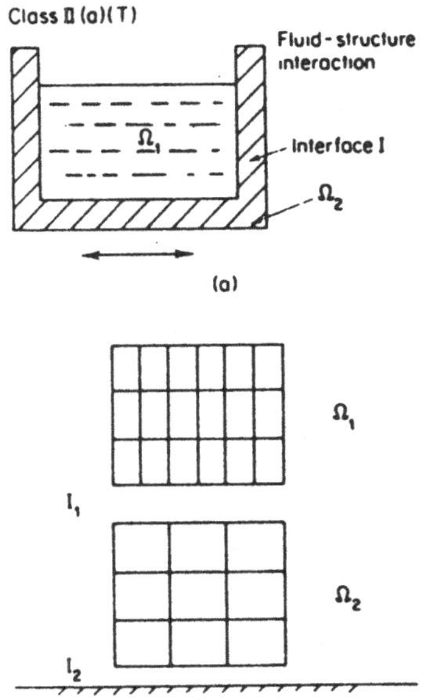

Fig. 2 Typical Class II problems.

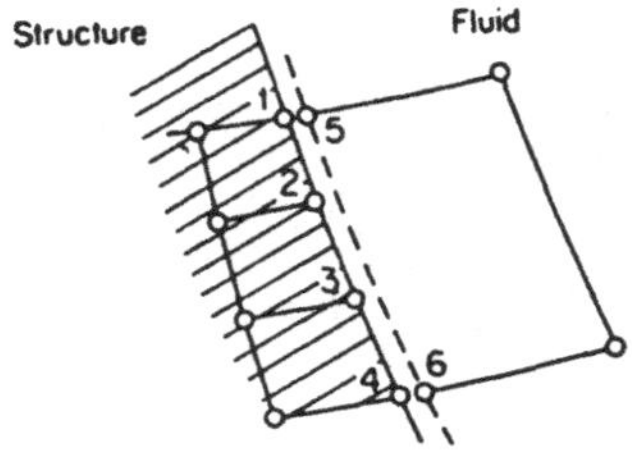

Fig. 3 A typical structure-fluid interface with a coarser mesh used for the fluid.

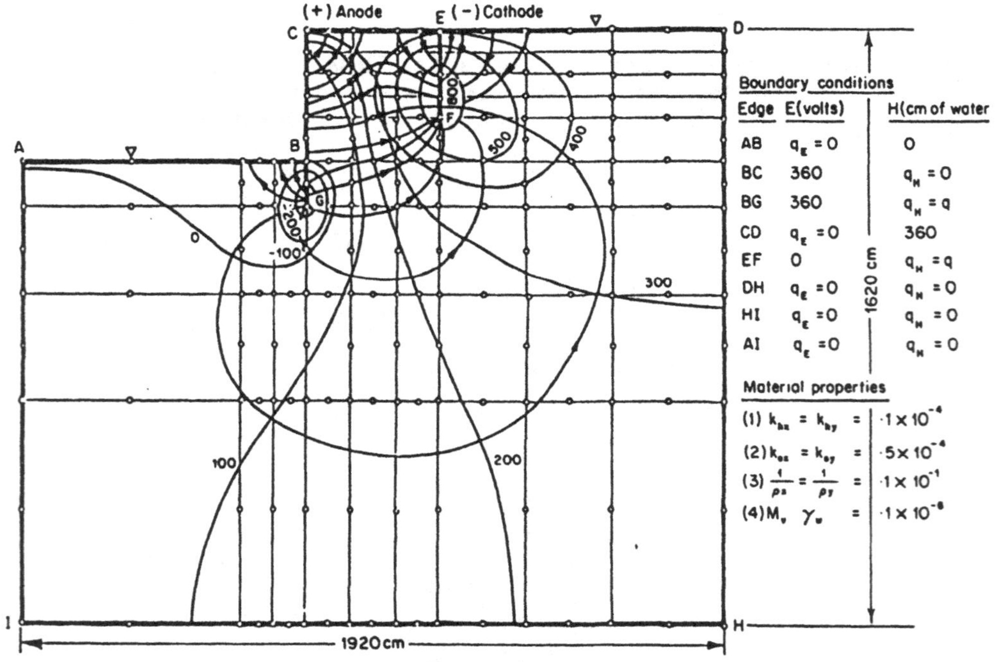

Fig. 4 Coupled electro-osmotic flow in an excavation: $\nabla^T(k_h \nabla H + k_e \nabla E) = X_{hh}\frac{\partial H}{\partial t} + X_{he}\frac{\partial E}{\partial t}$,

$$\nabla^T(g_h \nabla H + g_e \nabla E) = X_{eh}\frac{\partial H}{\partial t} + X_{ee}\frac{\partial E}{\partial t},$$

where H = hydrostatic head; E = electrical potential; k_h and k_e are permeabilities; X_i are storage coefficients and g_h and g_e are conductivities.

Material Properties

$K = 0.4302$ cal/cm.s. K

$\rho = 0.00275$ kg/cm^3

$C = 239.01$ cal/kg K

$\alpha_1 = 0.7314 \times 10^{-3}$ cal/cm^2 K

$\alpha_2 = 0.717 \times 10^{-2}$ cal/cm^2 K

σ_y (MN/m^2) $= \frac{1}{C_1} \left[(\frac{z}{C_2})^{1/C_3} + \sqrt{(\frac{z}{C_2})^{2/C_3} + 1} \right]$

$z = \bar{\dot{\varepsilon}} \, e^{C_4/RT}$ with

$C_1 = 0.01901$ m^2/MN

$C_2 = 7.92 \times 10^8$ sec^{-1}

$C_3 = 5.0$

$C_4 = 1.39 \times 10^5$ J/g mole

$R = 8.311$ J/g mole   Note: all dimensions in cm

$v_{tang} = v_{troll} = 28.73$ cm/s

$T = T_1$ $K\frac{\partial T}{\partial n} = \alpha_1(T-295)$ $K\frac{\partial T}{\partial n} = \alpha_2(T-322)$ $K\frac{\partial T}{\partial n} = \alpha_1(T-295)$

30.2

30.2

0.889

0.663

$v = 0$ $\frac{\partial T}{\partial n} = 0$

$\frac{\partial T}{\partial n} = 0$

x, u

2.25

3.665

2.25

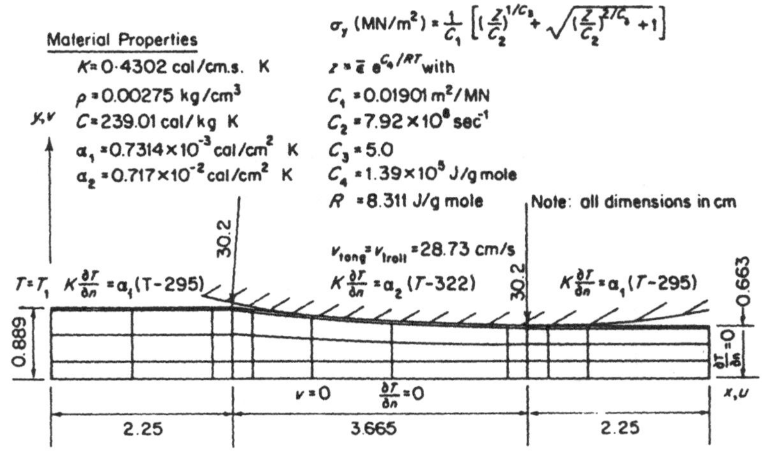

Geometrical configuration boundary conditions and finite element mesh used in hot rolling problem

(a)

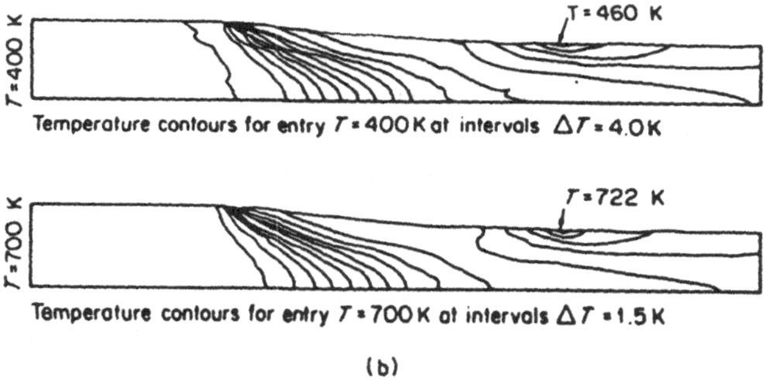

T = 400 K

T = 460 K

Temperature contours for entry $T = 400$ K at intervals $\Delta T = 4.0$ K

T = 700 K

T = 722 K

Temperature contours for entry $T = 700$ K at intervals $\Delta T = 1.5$ K

(b)

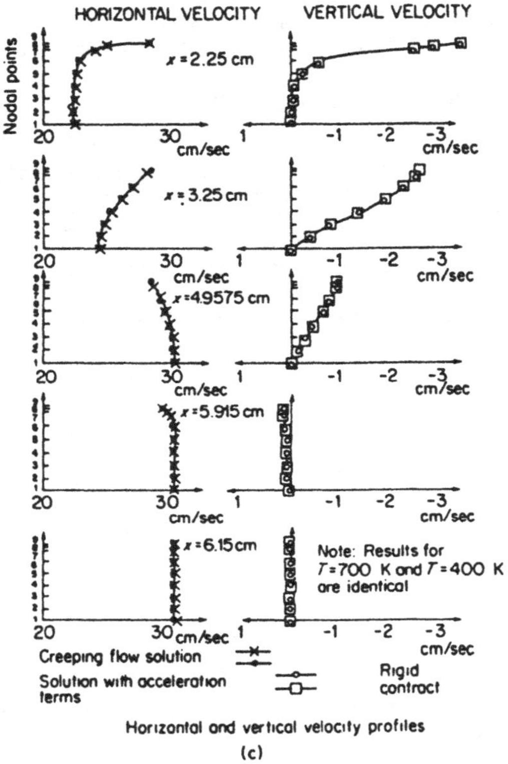

Fig. 5 Metal rolling process: thermally coupled flow.

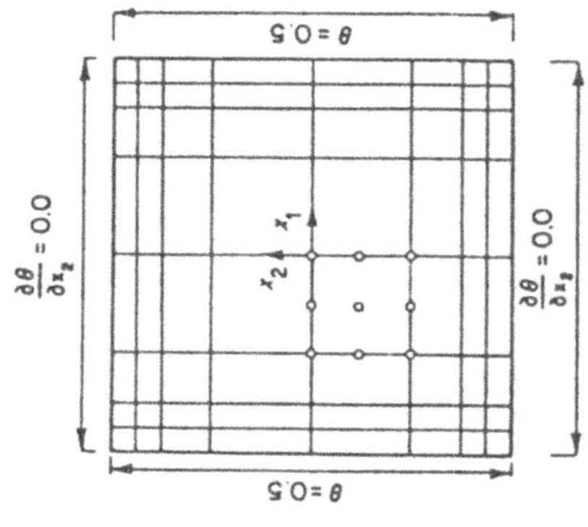

Problem geometry and computational grid

(a)

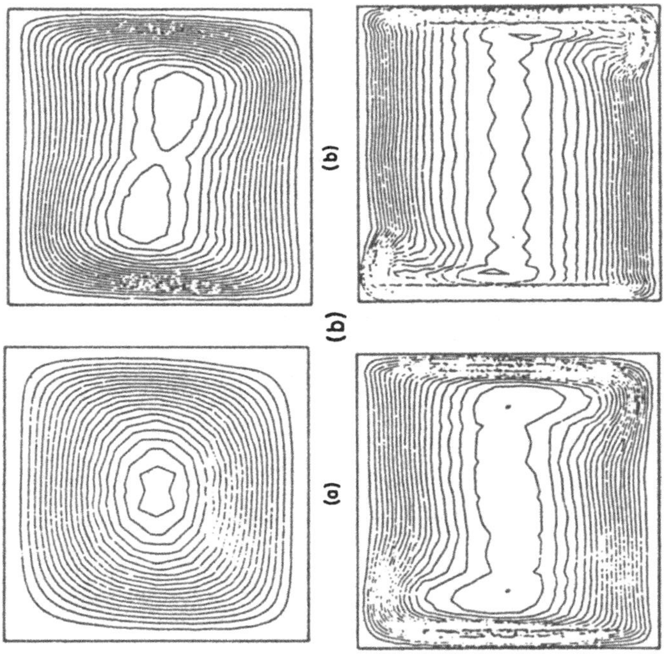

(b)

Stream function contours (a) Ra = 10^4, (b) Ra = 10^5, (c) Ra = 10^6, and (d) Ra = 10^7

Fig. 6 Gravity-induced thermal circulation in a heated cavity.

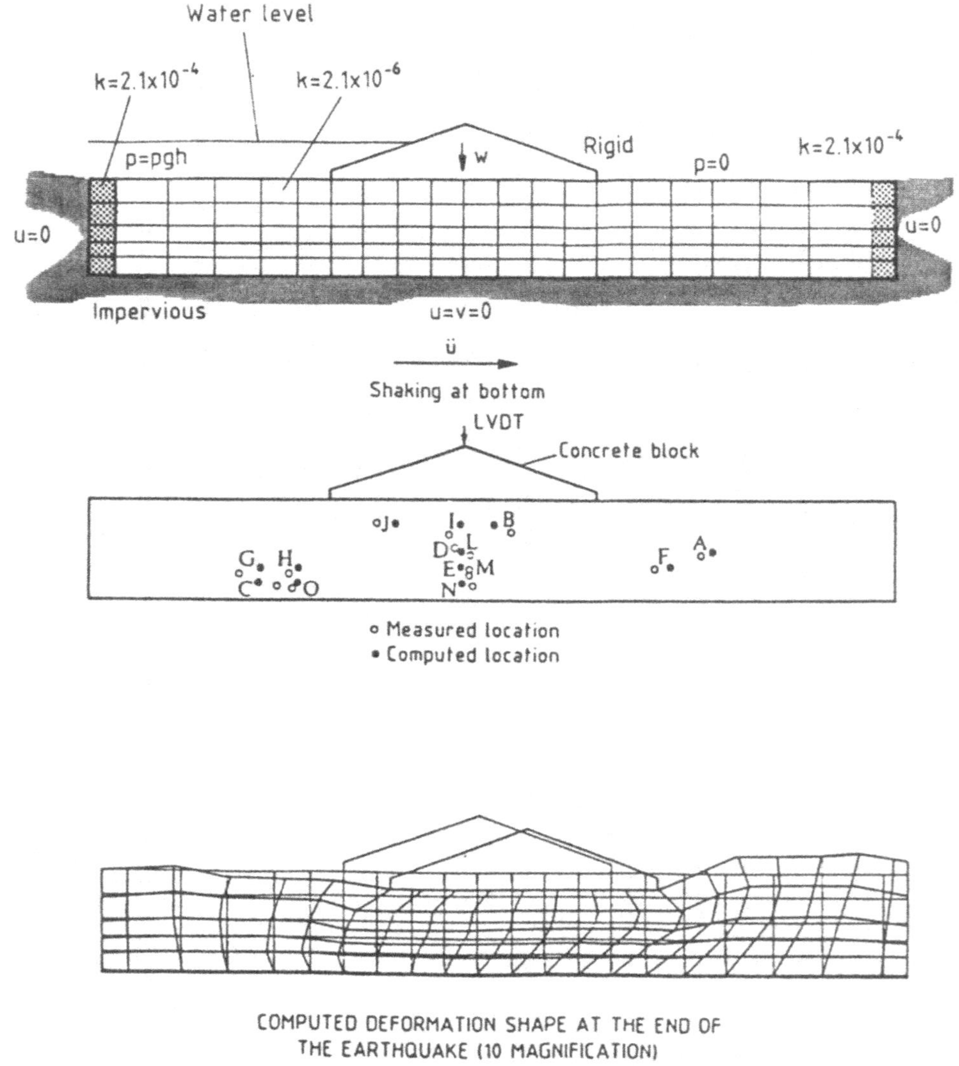

Fig. 7 Assessment of code by experiment. Swandyne (stag) vs centrifuge.

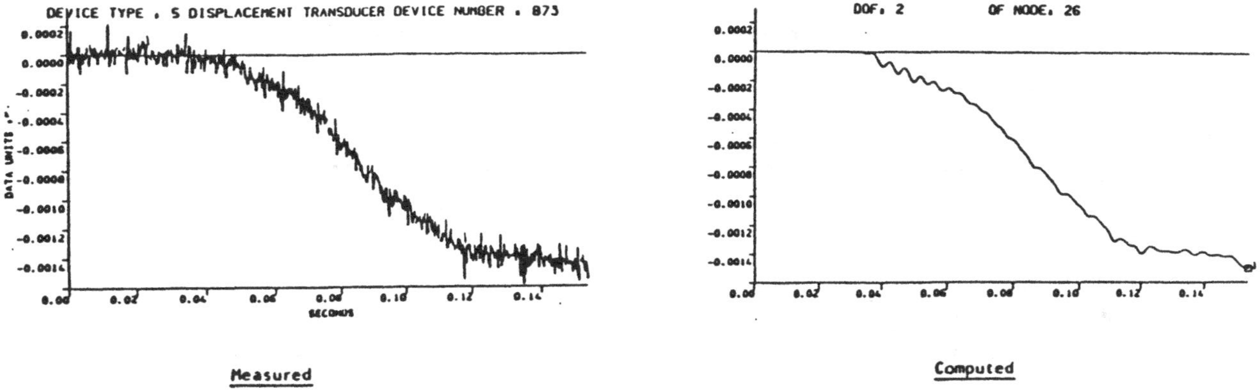

Fig. 8a Vertical displacement of the dyke.

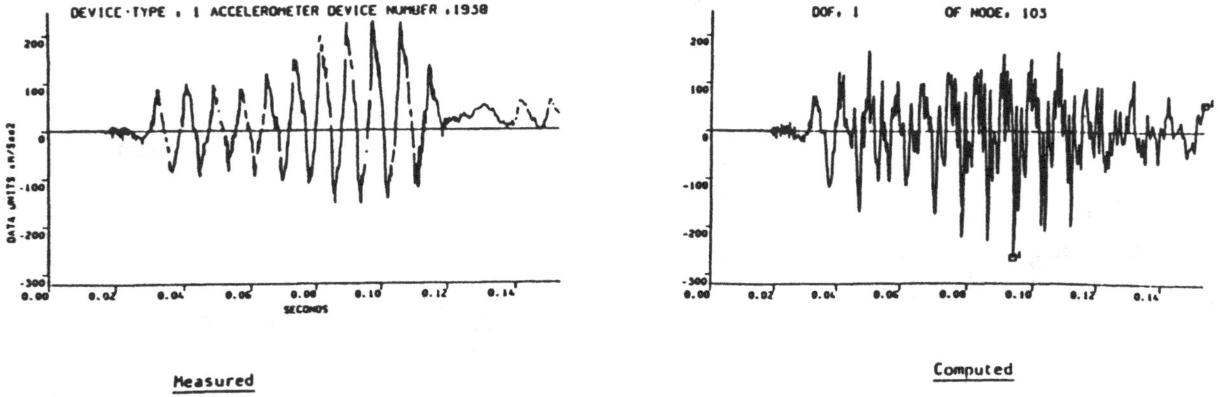

Fig. 8b Acceleration at point O.

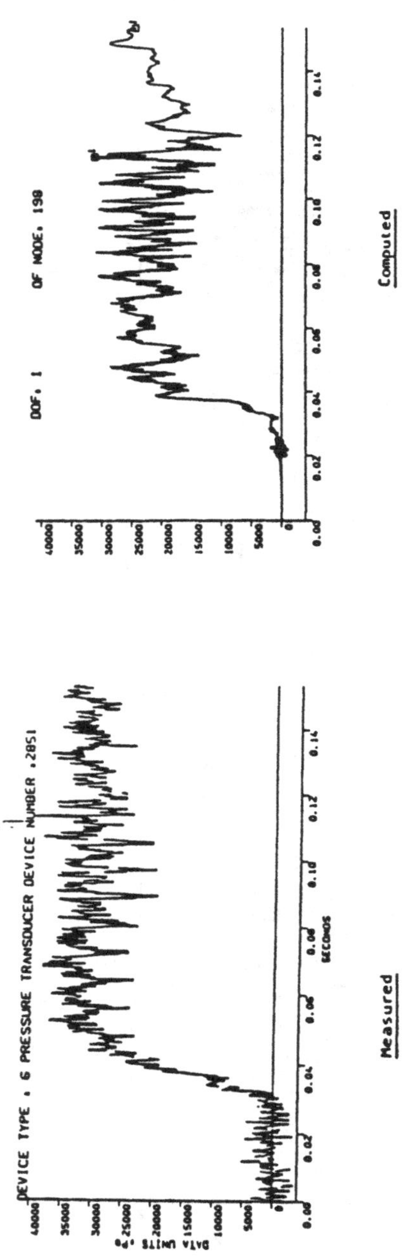

Fig. 8c Excess pore pressure at point A.

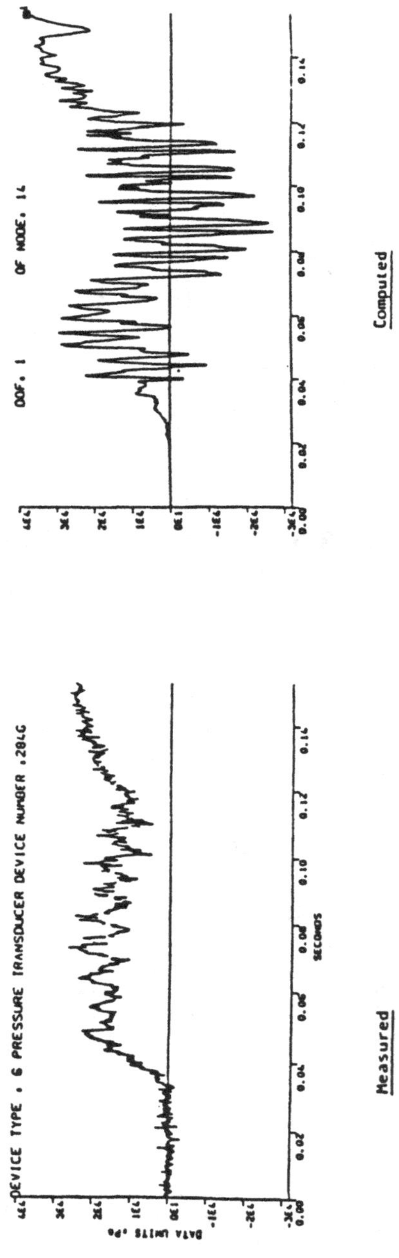

Fig. 8d Excess pore pressure at point B.

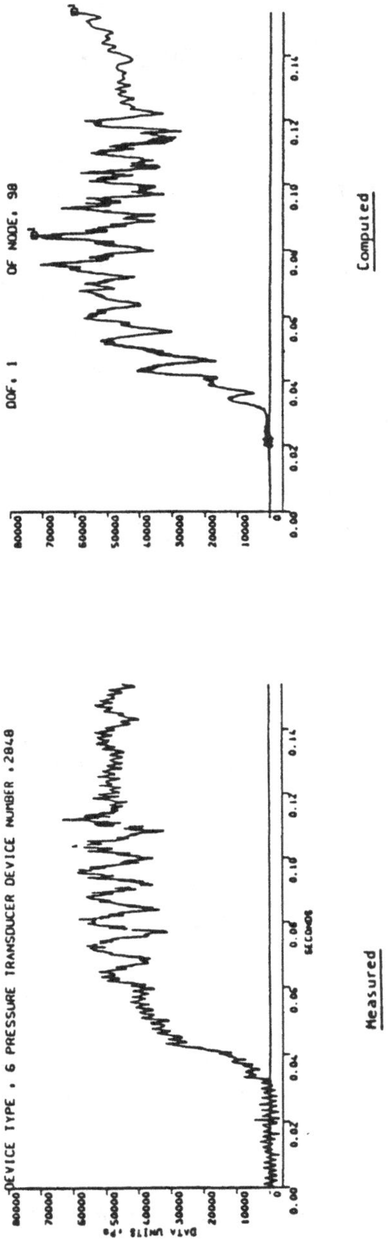

Fig. 8e Excess pore pressure at point C.

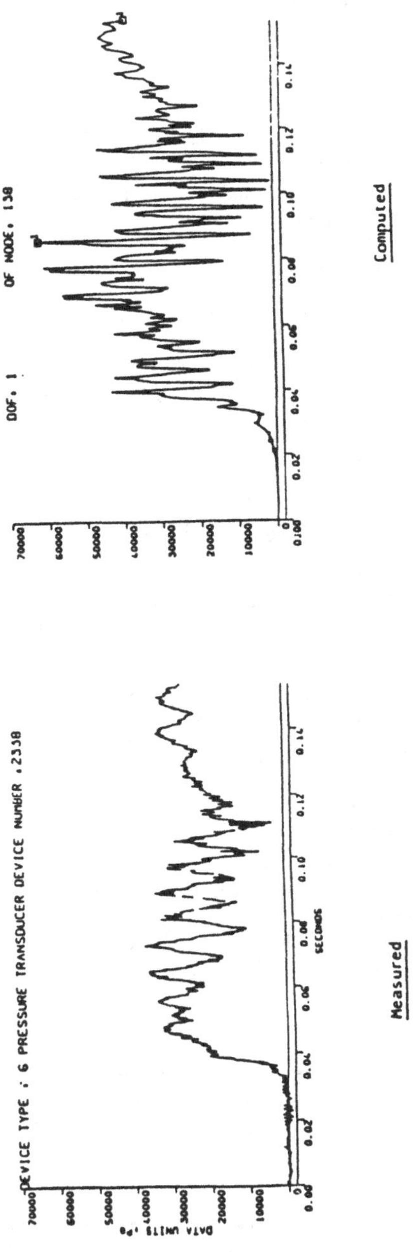

Fig. 8f Excess pore pressure at point D.

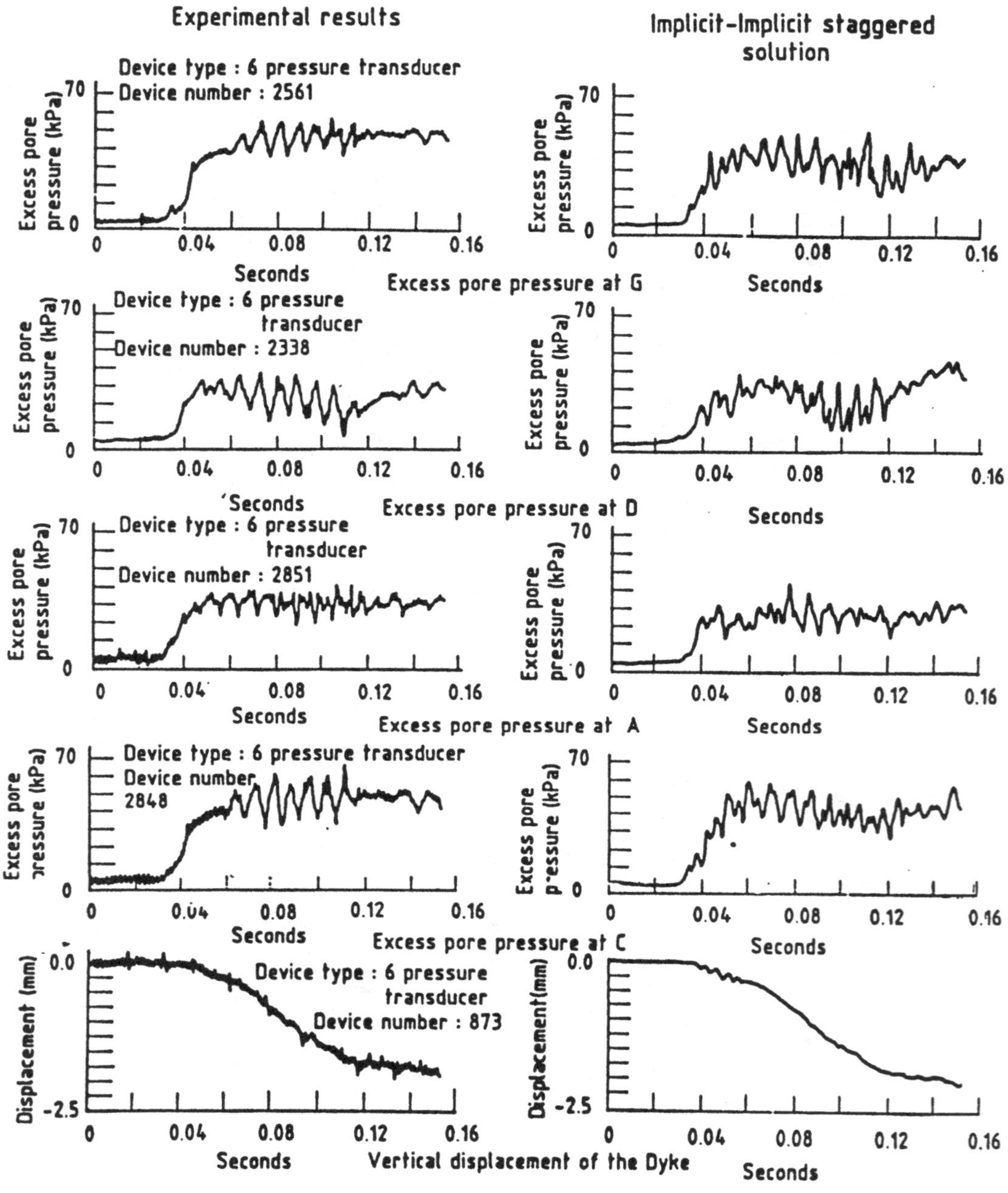

Fig. 9a Pore pressure and displacement. Computation vs experiment. (Short duration $0 < t < 0.16$ s)

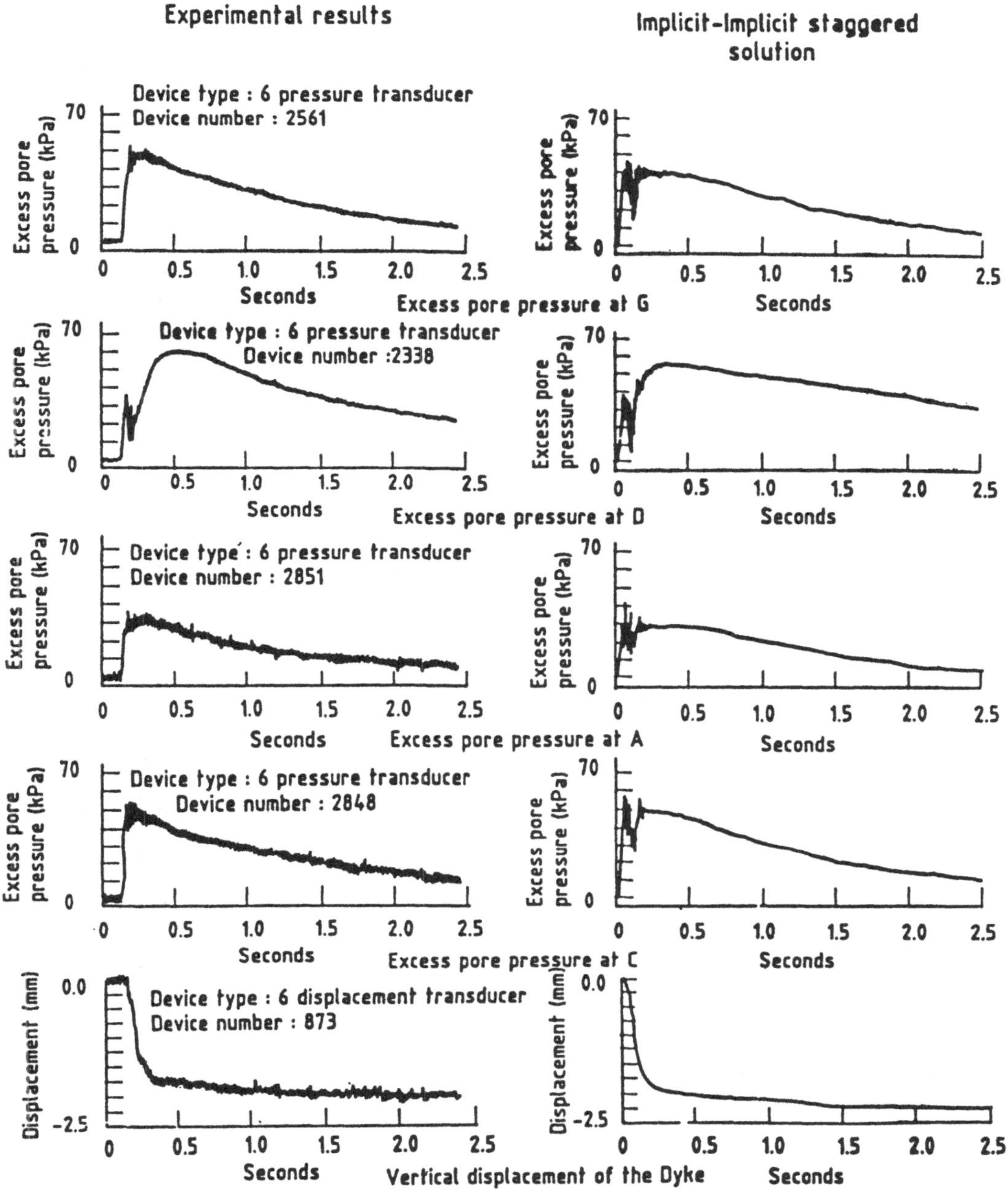

Fig. 9b Pore pressure and displacement. Computation vs experiment. (Long duration $0 < t < 2.5$ s)

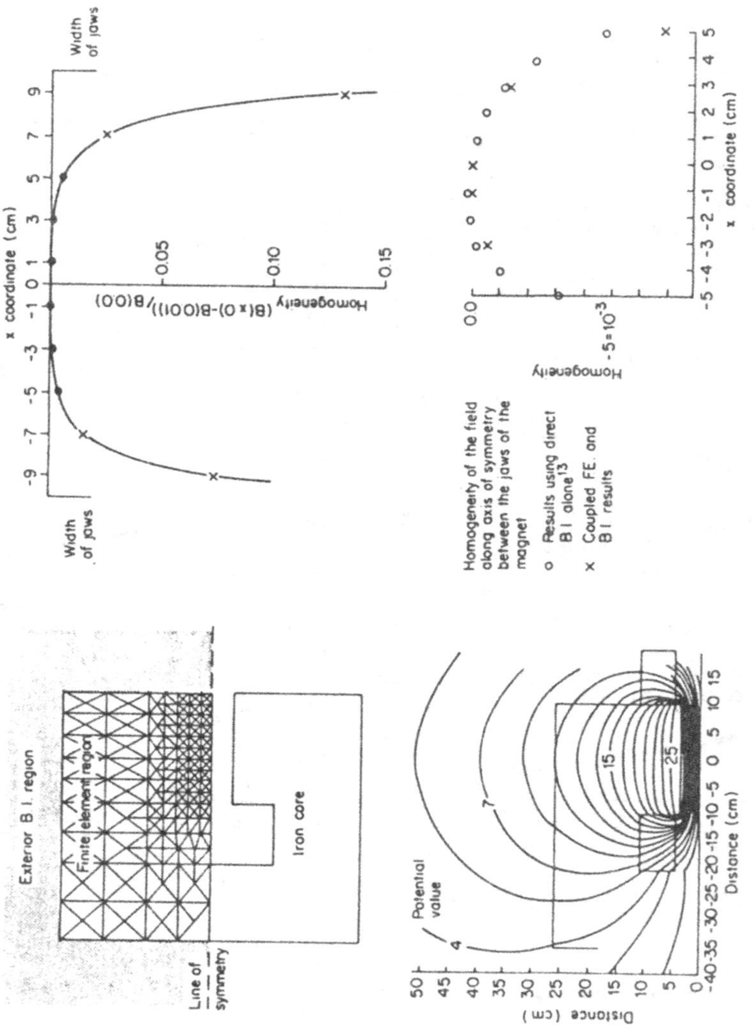

Fig. 10 Boundary integral and finite element solution coupled in solving an electromagnetic problem.

CHAPTER 5

GENERALIZED GALERKIN METHODS FOR CONVECTION-DOMINATED PROBLEMS

J. Donea
Joint Research Centre, Ispra, Italy

ABSTRACT

A brief survey is presented of generalized Galerkin methods for steady and unsteady convection-dominated problems. The particular areas upon which attention is focused are Petrov-Galerkin methods for steady problems and Taylor-Galerkin methods for evolutionary problems governed by hyperbolic equations. Also discussed are mixed advection-diffusion problems for which a splitting-up method of solution is suggested. Numerical examples are included to illustrate the performance of the Taylor-Galerkin method in the solution of linear and nonlinear hyperbolic problems.

1. INTRODUCTION

As is well-known, the Galerkin finite-element method and central-difference methods (which are intimately related) can be shown to possess the "best approximation" property when applied to symmetric spatial operators, like diffusion operators. That is, the difference between the numerical solution and the exact solution is minimized with respect to a certain norm.

The situation is fundamentally different in the case of purely convective transport. In fact, convection operators are non-symmetric and the self-adjointness, which is at the basis of the success of Galerkin and central-difference methods in pure diffusion cases, is therefore lost. The consequence is that numerical solutions to convection-dominated flows via the Galerkin finite-element method are frequently corrupted by physically unrealistic node-to-node oscillations, or wiggles, and these can only be removed by severe mesh and time-step refinements which clearly undermine the practical utility of the method.

Practitioners of finite-difference methods had soon realized that "upwind" differences can be used successfully to eliminate the oscillations caused by central differences in convection-dominated problems. However, the classical upwind difference scheme generally introduces an excessive numerical diffusion and consequently severely degrades the accuracy. This is the classical dilemma of upwind schemes: a competition between accuracy and stability. Nevertheless, the practical need to stabilize flow calculations has in the past necessitated the use of such procedures, accuracy degradation notwithstanding.

Recent finite-element efforts have revealed that stabilization of convection operators need not automatically engender accuracy loss. These efforts are briefly reviewed in the present chapter. In section 2, we discuss special Petrov-Galerkin methods which are adapted to deal with steady-state convection-dominated problems. Historically, these were the first problems to be tackled by finite-element developers and it turns out that some of the Petrov-Galerkin methods devised for steady-state situations can also be employed with success in the solution of transient problems. Then, in section 3, we consider generalized Galerkin methods that are specifically adapted to deal with purely transient situations. These include the so-called characteristic Petrov-Galerkin method and the Taylor-Galerkin method.

Only the main ideas which are at the basis of the generalized Galerkin methods for convection-dominated flows are illustrated herein. A more detailed discussion of these methods can be found in the references listed at the end of this chapter.

2. PETROV-GALERKIN METHODS FOR STEADY PROBLEMS

We will first consider a simple one-dimensional model problem which demonstrates the need for a generalization of the standard Galerkin finite-element method to accurately treat steady convection-dominated problems.

Our model equation is the linear, constant coefficient, one-dimensional convection-diffusion equation:

$$- au_x + k\, u_{xx} = 0 \qquad (1)$$

where a is the convective velocity and k a positive diffusion coefficient.

On a uniform mesh of linear elements with nodes numbered sequentially, the Galerkin approximation to Eq.(1) yields the following discrete equation at an interior node I:

$$(1-\gamma)u_{I+1} - 2u_I + (1+\gamma)u_{I-1} = 0 \qquad (2)$$

where: $\gamma = ah/2k$ (3)

is the element Péclet number and h denotes the length of an element. Eq.(2) indicates that the Galerkin finite element discretization of Eq.(1) using piecewise linear interpolations results in central difference approximations.

As any discrete equation, Eq.(2) possesses a truncation error. To analyse it, we expand Eq.(2) in Taylor series around u_I and use the original differential equation (1) to replace higher-order derivatives in terms of u_{xx}. This development indicates that the difference equation (2) is, in fact, equivalent to

$$-au_x + ku_{xx} + \frac{k}{2\gamma}\left(\frac{1}{\gamma}(\cosh 2\gamma - 1) - \sinh 2\gamma\right)u_{xx} = 0 \qquad (4)$$

Therefore, the second term of Eq.(4) is the exact truncation error of the discrete equation (2) and this error is in the form of a diffusion operator.

Now, it can be easily shown (see ref. [1]) that the function in Eq.(4) involving the element péclet number γ is systematically negative for all values of γ. We conclude, therefore, that the difference equation (2) derived from the Galerkin approximation to Eq.(1) solves a modified <u>underdiffuse</u> equation which we may write in the form

$$-a\, u_x + (k-k^*)u_{xx} = 0 \qquad (5)$$

where k is the physical diffusion coefficient and k^* the spurious artificial diffusion.

Thus, for values of the element Péclet number γ leading to $k^* > k$, we are, in fact, solving a modified problem with a negative global diffusion coefficient. This is clearly the cause of the numerical difficulties encountered in standard Galerkin, and central-difference solutions of steady-state convection-dominated problems.

Since the Galerkin approximation to the convection-diffusion equation (1) introduces a negative artificial diffusion, a justified remedy to improve the numerical solution is the use of an added diffusion, so as to effectively counterbalance the above spurious negative diffusion.

2.1 Upwind methods

In the finite difference context, diffusion can be added by the use of upwind differencing on the convective term. This consists of shifting the first derivative term in Eq.(1) in the upstream direction by writing

$$a(u_x)_I = \frac{a}{h}(u_I - u_{I-1}); \qquad a > 0 \tag{6}$$

However, in most circumstances upwind differencing as given by Eq.(6) yields numerical solutions which are very overdiffuse and this has been the basis for extensive criticism of upwind methods. Since central difference solutions are underdiffuse and upwind solutions overdiffuse, it appears that a linear combination of central and upwind differences represents the optimal solution method. As shown in ref. [1], an optimal upwind procedure can be formulated for the simple problem in Eq.(1) which yields nodally exact solutions for all values of the element Péclet number. Also shown in ref. [1] is that the optimal upwind method may be constructed by adding a proper amount of artificial diffusion to the central difference method. Nevertheless, problems of accuracy still remain in more complicated situations, such as those involving several space dimensions, or variable flow fields and source terms. See refs. [13-14] for a discussion of these problems.

Several techniques have been utilized to achieve the upwind effect in finite element formulations of steady convection-diffusion problems. As discussed in ref. [1], the key idea in achieving finite element methods of upwind type has been to replace the standard Galerkin method with a so-called Petrov-Galerkin finite-element method, in which the weighting function may be selected from a different class of functions than the approximate solution. The original ideas in this area were presented by the Dundee and Swansea research groups.

More recently, Hughes and co-workers have exploited the interpretation of upwind differences as central differences plus an artificial diffusion to construct upwind finite-element schemes by adding an appropriate artificial diffusion to the physical diffusion, while remaining in the conventional Galerkin finite element setting.

2.2 Generalized upwind finite element procedures

For the advection-diffusion problem in Eq.(1), with Dirichlet boundary conditions at $x = 0$ and $x = L$, Hughes and Brooks [2] replace the usual weak formulation

$$\int_0^L (w\, a\, u_x + w_x k\, u_x)\, dx = 0 \tag{7}$$

where w designates the weighting function, by the following weak statement

$$\int_0^L (w\, a\, u_x + w_x (k+\tilde{k})\, u_x)\, dx = 0 \tag{8}$$

where
$$\tilde{k} = \beta a h/2 \tag{9}$$

represents an artificial diffusion, the magnitude of which is governed by the free parameter β $(0 \leq \beta \leq 1)$. The value $\beta = 1$ corresponds to full upwind differencing as in Eq.(6).

In view of the definition (9) of the added diffusion, we note that the weak statement (8) can be rewritten in the form

$$\int_0^L \left[\left(w + \beta \frac{h}{2} w_x\right) a\, u_x + w_x\, k\, u_x\right] dx = 0 \tag{10}$$

Eq.(10) indicates that the added diffusion method can be regarded as a Petrov-Galerkin method in which a modified weighting function given by

$$\tilde{w} = w + \beta \frac{h}{2} w_x \tag{11}$$

is applied to the convective term. Notice that the above modified function is discontinuous at interelement boundaries.

Subsequently, Hughes and co-workers extended the above method to deal with source terms and several space dimensions (see refs. [2,9] for details). They first pointed out that when a source term is present in the governing equation, this term must be discretized, like the convective term, using the modified weighting function in Eq.(11). If this is not done, very poor results would again be obtained. Finally, for the extension of the added diffusion method to deal with multidimensional problems, Hughes and Brooks proposed to construct the artificial diffusion operator in tensorial form to act only in the flow direction and not transversely. The scalar artificial diffusion in Eq.(9) is now replaced by a tensor diffusivity,

$$\tilde{k}_{ij} = \bar{k}\, a_i\, a_j / \|a\|^2, \tag{12}$$

where $\bar{k}$ is a scalar artificial diffusivity, and a_i is the flow velocity component along the coordinate direction x_i. It is readily noted that Eq.(12) represents a diffusivity acting only in the direction of the flow and not transversely. This is the so-called streamline-upwind finite-element method, which was successfully applied by Hughes and co-workers to the solution of convection-diffusion and incompressible Navier-Stokes equations. The streamline-upwind method can be regarded as a Petrov-Galerkin method in which modified weighting functions given by

$$\tilde{w} = w + \left(\frac{\bar{k}}{\|a\|^2}\right) a \cdot \nabla w, \qquad (13)$$

are applied consistently to the convective term, and the source term in the governing equation. The method was also applied with success to transient problems; in this case, the modified function (13) is applied also in the weighting of the time derivative term.

3. GENERALIZED GALERKIN METHODS FOR TRANSIENT PROBLEMS

The previous section was concerned with a discussion of generalized Galerkin methods for steady problems describing convection and diffusion. Over the last few years, efforts have also been devoted to the development of generalized Galerkin methods specifically adapted to the treatment of unsteady problems.

For simplicity, we first consider purely convective problems governed by first-order hyperbolic equations or systems of equations. Methods for mixed convection-diffusion problems will be reviewed at a later stage.

As a simple example of purely convective problems, consider the scalar convection equation in one dimension

$$u_t + a\,u_x = 0, \quad a > 0. \qquad (14)$$

On a uniform mesh of linear elements with nodes numbered sequentially, the Galerkin approximation to Eq.(14) yields the following semidiscrete equation at an interior node I:

$$\dot{u}_I + \frac{1}{6}\left(\dot{u}_{I-1} - 2\dot{u}_I + \dot{u}_{I+1}\right) + \frac{a}{2h}\left(u_{I+1} - u_{I-1}\right) = 0 \qquad (15)$$

Here, a superposed dot indicates a time derivative. Note that the Galerkin method introduces a supplementary term, with respect to finite differences, in the discretization of the time derivative term. This term yields the so-called consistent mass matrix of the finite-element method which, in the present case, offers the advantage of improving the spatial accuracy to fourth order (see ref. [4] for details). This represents indeed a significant gain with respect to the second-order accuracy achieved with standard finite differences. However, the system of ordinary differential equations (15) still has to be integrated forward in time to produce the transient response, and the fourth-order spatial accuracy can be quickly eroded if the numerical time-integration procedure is not of comparable accuracy.

3.1 Petrov-Galerkin methods

Morton and his colleagues were the first to point out that the key problem in solving advection problems with finite elements was to achieve a proper coupling between space discretization and time discretization. They clearly pointed out that the classical, second-order accurate, time stepping algorithms (i.e. the Lax-Wendroff, leap-frog, and Crank-Nicolson methods) properly combine with piecewise linear finite elements in advection problems only for small values of the time step Δt. On the contrary, as the Courant number

$$C = a\,\Delta t/h \tag{16}$$

increases, the numerical phase error does not decrease uniformly at all wavelengths, so that the optimal stability limit for explicit schemes ($C \leq 1$) and the so-called unit CFL property (exact nodal solutions for $C = 1$ on a uniform mesh) are not achieved with the conventional Galerkin method, while virtually all practical finite-difference schemes do exhibit such optimal properties.

Morton and Parrott [4] considered whether simple weighting functions in a Petrov-Galerkin formulation could be derived, which produce exact answers on a uniform mesh of linear elements when $C = 1$. They found such weighting functions, which we denote by χ, for several time-stepping algorithms (Euler, leap-frog, Crank-Nicolson). Then, to match the high accuracy of the usual Galerkin method as $C \rightarrow 0$ with the property of the new functions χ which give exact answers at $C = 1$, Morton and Parrott advocate the use of test functions given by the linear combination

$$\tilde{w} = (1-\nu)w + \nu\chi \tag{17}$$

where w represents the standard Galerkin weighting functions and ν is a free parameter. However, the test functions (17) do not retain all the valuable conservation properties associated with the pure Galerkin formulation, and the generalization of the above Petrov-Galerkin schemes to multidimensional and nonlinear problems faces some difficulties. An approach more closely based on the characteristics was subsequently elaborated upon by Morton [6] and successfully applied to the solution of the Euler equations of inviscid gas dynamics.

Techniques combining the method of characteristics with finite-element procedures were also used with success for the Navier-Stokes equations (see, for example, Benqué et al. [15], Bercovier et al. [16] and Douglas and Russell [17]).

3.2 Taylor-Galerkin method

An alternative approach was taken by the present author with the introduction of the Taylor-Galerkin method as a means of generating high-order accurate time-stepping schemes to be coupled to the high spatial resolution attainable with the conventional Galerkin finite-element formulation. The concepts which lie at the basis of the Taylor-Galerkin method are illustrated in detail in ref. [5] and an analysis of the performance of the resulting schemes is presented in ref. [8]. Here, only a brief description of the method will be given.

In contrast with the usual procedure in which spatial discretization is performed before time discretization, the Taylor-Galerkin method considers time discretization before spatial discretization which is performed using the conventional Galerkin finite-element method.

As a simple example for illustrating the Taylor-Galerkin method, we consider the scalar convection equation (14) in conjunction with explicit Euler time stepping. We first leave the spatial variable x continuous and discretize only the time by means of the following Taylor series:

$$u^{n+1} = u^n + \Delta t\, u_t^n + \frac{1}{2}\Delta t^2\, u_{tt}^n + \frac{1}{6}\Delta t^3\, u_{ttt}^n + O(\Delta t^4) \tag{18}$$

which yields

$$(u^{n+1}-u^n)/\Delta t = u_t^n + \frac{1}{2}\Delta t\, u_{tt}^n + \frac{1}{6}\Delta t^2\, u_{ttt}^n + O(\Delta t^3) \tag{19}$$

Now, Eq.(14) and its successive differentiations with respect to time indicate that

$$\begin{aligned} u_t^n &= -a\, u_x^n \\ u_{tt}^n &= -a\, u_{xt}^n = a^2\, u_{xx}^n \\ u_{ttt}^n &= a^2 (u_t)_{xx}^n = a^2 \left(\frac{u^{n+1}-u^n}{\Delta t}\right)_{xx} + O(\Delta t) \end{aligned} \tag{20}$$

so that Eq.(19) can be rewritten in the form

$$\left(1 - \frac{a^2\Delta t^2}{6}\partial_{xx}\right)\left(\frac{u^{n+1}-u^n}{\Delta t}\right) = -a\, u_x^n + \frac{a^2\Delta t}{2}u_{xx}^n + O(\Delta t^3) \tag{21}$$

which, therefore, represents a generalized third-order accurate Euler time-stepping method. To obtain a fully discrete equation, we apply the standard Galerkin formulation to Eq.(21). In the case of a uniform mesh of linear elements, the following discrete equation is obtained at an interior node I:

$$\left[1 + \frac{1}{6}\left(1-C^2\right)\delta^2\right]\left(u_I^{n+1}-u_I^n\right) = -\frac{C}{2}\left(u_{I+1}^n - u_{I-1}^n\right) + \frac{1}{2}C^2\left(u_{I+1}^n - 2u_I^n + u_{I-1}^n\right) \tag{22}$$

where C is the Courant number defined in Eq.(16).

The following points should be stressed:

1) If the Taylor series (19) is limited to $O(\Delta t^2)$ and a lumped (diagonal) mass matrix is employed in (22), the familiar, second-order accurate, Lax-Wendroff method is obtained.
2) In spite of the appearance, the last term in Eq.(22) is not to be thought of as an additional artificial diffusion term, but rather as part of the improved difference approximation to u_t.
3) The third-order derivative term in Eq.(20) has purposely been expressed in a mixed spatial-temporal form, for the presence in the governing equation of a third-order spatial derivative would prevent

us from employing C^0 finite elements for spatial discretization, whereas the adopted mixed form simply leads to modify the usual mass matrix, which nevertheless remains symmetric.

4) The stability condition for the generalized Euler method in Eq.(22) is $C \leqq 1$ and the scheme possesses the unit CFL property, i.e. signals are propagated without distortion when the characteristics pass through the nodes. Moreover, as shown in [5,8], the scheme is found to exhibit particularly high phase-accuracy with minimal numerical damping.

In ref. [5], generalized leap-frog and Crank-Nicolson methods are developed, which possess a fourth-order accuracy. Another interesting characteristic of the Taylor-Galerkin method is that it is readily extended to deal with multidimensional problems. In two and three dimensions, the added diffusion term as a result of the improved temporal accuracy is found to automatically possess the tensorial structure which was advocated by Hughes and Brooks in their streamline-upwind Petrov-Galerkin formulation (cf. Eq.(12)). This may be seen by considering the scalar convection equation

$$u_t + \mathbf{V}\cdot\nabla u = 0 \tag{23}$$

where the velocity field V is assumed to be time-independent and divergence-free. In these conditions, the second time derivative reads

$$u_{tt} = -\nabla\cdot(u_t \mathbf{V}) = \nabla\cdot[(\mathbf{V}\cdot\nabla u)\mathbf{V}] \tag{24}$$

and the last term in (24) clearly indicates that the diffusion term added by the Taylor-Galerkin method does possess a tensorial structure similar to that in the streamline-upwind Petrov-Galerkin formulation.

Finally, the Taylor-Galerkin method can be applied in the solution of nonlinear advection problems and to mixed advection-diffusion problems, as will now be briefly illustrated. See [7] for a complete discussion.

3.3 Nonlinear advection

As a model equation for illustrating the application of the Taylor-Galerkin method to nonlinear advection problems, consider the scalar conservation law in one dimension

$$u_t + \partial_x f(u) = 0, \tag{25}$$

or

$$u_t + a(u)u_x = 0, \tag{26}$$

where

$$a(u) = \partial f(u)/\partial u. \tag{27}$$

Following the procedure used in Eqs.(19-(20) to generate a generalized Euler time-stepping method, we obtain from the governing equation (26) and its time differentiation:

$$u_t = - a\, u_x \tag{28}$$

and

$$u_{tt} = - a_t u_x - a\, u_{xt} \tag{29}$$

which, noting that

$$a_t = a_u u_t = -a\, a_u\, u_x = -a\, a_x \tag{30}$$

and

$$u_{xt} = - a_x u_x - a\, u_{xx} \tag{31}$$

may be transformed into

$$u_{tt} = 2a\, a_x u_x + a^2 u_{xx} = \partial_x (a^2 u_x) \tag{32}$$

Similarly, by time differentiation of Eq.(32) and use of Eq.(30) we find

$$u_{ttt} = - \partial_x (2a^2 a_x u_x) + \partial_x (a^2 u_{tx}) \tag{33}$$

We can now replace the first, second and third time derivatives in Eq. (19) by expressions (28, (32) and (33), respectively, and obtain, upon approximation of u_{tx} in (33) by $\partial_x[(u^{n+1}-u^n)/\Delta t]$, the following generalized, third-order accurate Euler time-stepping method for the scalar conservation law (26):

$$\left[1 - \partial_x \left(\frac{a^2 \Delta t^2}{6} \partial_x \right)\right] \left(\frac{u^{n+1} - u^n}{\Delta t} \right) = -(a\, u_x)^n + \frac{1}{2} \Delta t \partial_x \left[a^2 \left(1 - \frac{2}{3} \Delta t a_x \right) u_x \right]^n \tag{34}$$

Selmin et al. [7] have further generalized the above method to treat in one and two space dimensions the Euler equations governing the transient flow of inviscid compressible fluids. However, the above Taylor-Galerkin procedure becomes quite cumbersome and time-consuming in application to systems of nonlinear hyperbolic equations. In this case, it is more convenient to resort to two-step Taylor-Galerkin methods which represent the generalization to finite elements of the two-step Lax-Wendroff methods introduced in the finite difference literature by Richtmyer [18] and further refined by Burstein [19] and others.

To illustrate the construction of two-step Taylor-Galerkin methods, consider the Euler equations written in the form

$$\mathbf{W}_t + \nabla \cdot \mathbf{F} = 0, \tag{35}$$

where $\mathbf{W}$ is the vector of conservation variables (density, momentum components, total energy) and $\mathbf{F}(\mathbf{W})$ are corresponding flux vectors. A two-step version of the second-order Taylor-Galerkin scheme may be obtained as follows for the case of linear isoparametric elements in which both $\mathbf{W}$ and $\mathbf{F}$ are locally approximated in terms of nodal values as

$$\mathbf{W} = N_I \mathbf{W}_I; \quad \mathbf{F} = N_I \mathbf{F}_I \tag{36}$$

where the N_I's are the multilinear shape functions.

STEP 1: For each element, calculate a constant element vector $\mathbf{W}_e^{n+\frac{1}{2}}$ from

$$A_e \mathbf{W}_e^{n+\frac{1}{2}} = \sum_{I=1}^{N^*} \left(\int_{\Omega_e} N_I d\Omega \right) \mathbf{W}_I^n - \frac{\Delta t}{2} \int_{\Omega_e} \nabla \cdot \mathbf{F}^n d\Omega \tag{37}$$

where A_e denotes the volume of element e and N^* indicates the number of nodes in the element ($N^* = 4$ in 2D; $N^* = 8$ in 3D).

STEP 2: For each node I, calculate $\mathbf{W}_I^{n+1}$ by solving the following system of equations:

$$\sum_{J=1}^{N} \left(\int_{\Omega} N_I N_J d\Omega \right) \left(\mathbf{W}_J^{n+1} - \mathbf{W}_J^n \right) \Big/ \Delta t =$$

$$= \int_{\Omega} \nabla N_I \cdot \mathbf{F}_e^{n+\frac{1}{2}} d\Omega - \int_{\partial\Omega} N_I \left(\mathbf{F}_e^{n+\frac{1}{2}} - \mathbf{F}_e^n \right) \cdot \mathbf{n}\, ds - \int_{\partial\Omega} N_I \mathbf{F}^n \cdot \mathbf{n}\, ds \tag{38}$$

Here, N indicates the total number of nodes, $\mathbf{F}_e^{n+\frac{1}{2}}$ denotes the flux vector evaluated using the constant element vector $\mathbf{W}_e^{n+\frac{1}{2}}$ emanating from Step 1 and $\mathbf{F}_e^n$ represents the elementwise averaged value of the flux at time $n\Delta t$.

Equations (37) and (38) define a two-step Taylor-Galerkin method with second-order temporal accuracy. The method was originally proposed by Oden et al. [20] and used for solving high-speed compressible flows. Artificial diffusion was added in [20] in order to stabilize the scheme in the presence of discontinuities. One should also note that hour-glassing modes may develop due to the elementwise averaged fluxes used in the second step of the calculation. To avoid this problem, we are currently investigating an alternative formulation of Step 1 in which $\mathbf{W}^{n+\frac{1}{2}}$ is evaluated at the element Gauss points. Such a formulation has the further advantage of extending the applicability of the two-step Taylor-Galerkin method to higher-order elements.

3.4 Mixed advection-diffusion problems

In view of the different mathematical characters of diffusion and convection operators, it appears appropriate to treat unsteady advection-diffusion problems by a splitting-up method in which the mixed problem is decomposed into a pure convection problem followed by a pure diffusion problem. By such a decomposition, convection can be treated by the

Taylor-Galerkin method discussed above and another appropriate algorithm may be used for the diffusion phase.

Consider, for example, the solution of Burgers' equation

$$u_t = - \partial_x\left(\frac{1}{2} u^2\right) + k\, u_{xx} \tag{39}$$

Numerically, the convection phase can be treated by the third-order Euler method in Eq.(34) with, for the present case, $a = u$. This gives intermediate values, $\tilde{u}$, which are then used for the diffusion phase. This last phase can, for example, be based on the following second-order Euler method:

$$\left(1 - \frac{k\Delta t}{2}\, \partial_x^2\right)\left(\frac{u^{n+1} - \tilde{u}}{\Delta t}\right) = k\, \tilde{u}_{xx} \tag{40}$$

As shown in refs. [10,11], splitting-up procedures of the above type are easily generalized to treat more complex equations such as, for instance, the incompressible Navier-Stokes equations in the velocity-pressure formulation. In these equations, only the incompressibility condition needs to be treated implicitly, while the advection and viscosity terms can be treated explicitly. One is then led to a three-stage splitting-up procedure dealing successively with the advection terms, the diffusion terms, and the pressure calculation. Algorithms of this type are very attractive computationally and can be applied as well in the solution of compressible flow problems.

4. ILLUSTRATIVE EXAMPLES

Four test problems are presented to illustrate the performance of the Taylor-Galerkin method in the solution of linear and nonlinear hyperbolic problems.

4.1 Propagation of a cosine profile

Consider the advection problem over the spatial interval [0,1] defined by the initial and boundary conditions [8]

$$u(x,0) = \begin{cases} \frac{1}{2}\{1+\cos[\pi(x-x_o)/\sigma]\} & \text{if } |x-x_o| \leq \sigma \\ 0 & \text{if } |x-x_o| > \sigma \end{cases}$$

$$u(0,t) = 0, \quad t \geq 0,$$

with $x_o = 0.2$ and $\sigma = 0.12$. The exact solution of Eq.(14) with $a = 1$ corresponds to the translation to the right of the initial profile with a unit velocity. Fig. 1 compares the numerical solutions obtained using a uniform mesh of 50 linear elements and different values of the Courant number C with the exact solution at $t = 0.6$. As may be seen, the Lax-Wendroff finite-difference scheme exhibits a significant phase error at small values of the Courant number. By contrast, the Lax-Wendroff finite element scheme (with a consistent mass matrix) and the third-order Taylor-Galerkin scheme (21) exhibit a rather uniform phase accuracy. Note, however, that the Lax-Wendroff finite-element scheme cannot be operated when $C^2 > 1/3$, while the Taylor-Galerkin scheme is stable for $C \leq 1$.

4.2 Rotating cone problem

Consider now the advection of a cosine hill in a pure rotation velocity field. The initial condition is [8]:

$$u_o(x_1,x_2) = \begin{cases} \frac{1}{4}(1+\cos \pi X_1)(1+\cos \pi X_2) & \text{if } X_1^2 + X_2^2 \leq 1 \\ 0 & \text{if } X_1^2 + X_2^2 \geq 1 \end{cases}$$

where $\mathbf{X} = (\mathbf{x}-\mathbf{x}_o)/\sigma$, $\mathbf{x}_o$ and σ being the initial position of the centre and the radius of the cosine hill.

A uniform mesh of 30x30 bilinear elements over the unit square has been employed in the calculations. The numerical solutions for the case $\mathbf{x}_o = \left(\frac{1}{6}, \frac{1}{6}\right)$ and $\sigma = 0.2$ are shown in Fig. 2 after a complete revolution for two time-step values, $\Delta t = 2\pi/200$ and $\Delta t = 2\pi/120$. The superiority of the Taylor-Galerkin (TG) scheme with respect to the Lax-Wendroff finite-difference (FD) and finite-element (FE) schemes is again clearly apparent.

4.3 Sod shock tube problem

Consider now the solution of a Riemann problem for a perfect gas - the well-known Sod shock tube problem. The calculations have been performed on a uniform mesh using a time-step corresponding to a CFL condition number of 0.9 for the explicit, two-step, Lax-Wendroff finite-difference scheme and the Taylor-Galerkin scheme, whereas a CFL number of 0.45 was employed for the second-order finite-element scheme. No explicit artificial viscosity has been used in the calculations. The results in Fig. 3 (extracted from [8]) show the superiority of the finite-element schemes which manifests itself by a better overall accuracy, a much reduced amplitude of non-physical oscillations and a sharper representation of the discontinuities.

4.4 Supersonic flow in a channel with a step

As a last example, we consider the transient supersonic (Mach 3) flow in a channel with a step. The spatial discretization is shown in Fig. 4, together with the computed pressure contours at several times. The present Taylor-Galerkin solution was found to be in good agreement with the reference solution produced by Woodward and Colella [12].

REFERENCES

1. J. Donea, Recent advances in computational methods for steady and transient transport problems, Nucl. Eng. Design, 80 (1984) 141-162.
2. T.J.R. Hughes and A.N. Brooks, A multi-dimensional upwind scheme

with no crosswind diffusion, in "Finite Element Methods for Convection Dominated Flows, T.J.R. Hughes, Ed., AMD - Vol.34, ASME, New York, 1979.

3. I. Christies, D.F. Griffiths, A.R. Mitchell and O.C. Zienkiewicz, Finite element methods for second order differential equations with significant first derivatives, Int. J. Num. Meths. Engng., 10 (1976) 1389-1396.
4. K.W. Morton and A.K. Parrott, Generalized Galerkin methods for first-order hyperbolic equations, J. Comput. Phys., 36 (1980) 249-270.
5. J. Donea, A Taylor-Galerkin method for convective transport problems, Int. J. Num. Meths. Engng., 20 (1984) 101-119.
6. K.W. Morton, Generalised Galerkin methods for hyperbolic problems, Comput. Meths. Appl. Mech. Engng., 52 (1985) 847-871.
7. V. Selmin, J. Donea and L. Quartapelle, Finite element methods for non-linear advection, Comput. Meths. Appl. Mech. Engng., 52 (1985) 817-845.
8. J. Donea, L. Quartapelle and V. Selmin, An analysis of time discretization in the finite element solution of hyperbolic problems, J. Comput. Phys., 70 (1987) 463-499.
9. T.J.R. Hughes and M. Mallet, A new finite element formulation for computational fluid dynamics: III. The generalized streamline operator for multidimensional advective-diffusive systems, Comput. Meths. Appl. Mech. Engng., 58 (1986) 305-328.
10. J. Donea, S. Giuliani, H. Laval and L. Quartapelle, Time-accurate solution of advection-diffusion problems by finite elements, Comput. Meths. Appl. Mech. Engng., 45 (1984) 123-145.
11. J. Donea, S. Giuliani, H. Laval and L. Quartapelle, Finite element solution of the unsteady Navier-Stokes equations by a fractional step method, Comput. Meths. Appl. Mech. Engng., 30 (1982) 53-73.
12. P. Woodward and P. Colella, The numerical simulation of two-dimensional fluid flow with strong shocks, J. Comput. Phys., 54 (1984) 115-173.
13. W. Shyy, A study of finite difference approximations to steady-state, convection-dominated flow problems, J. Comput. Physics, 57 (1985) 415-438.
14. I. Christie, Upwind compact finite difference schemes, J. Comput. Physics, 59 (1985) 353-368.
15. J.P. Benqué, G. Labadie and J. Ronat, A new finite element method for the Navier-Stokes equations coupled with a temperature equation, Proc. 4th Int. Conf. on Finite Element Methods in Flow Problems (Ed. Tadahiko Kavi) (1982) 295-301.
16. M. Bercovier, O. Pironneau, Y. Hasbani and E. Livne, Characteristic and finite element methods applied to the equations of fluids, Proc. MAFELAP 1981 Conf. (Ed. J.R. Whiteman), Academic Press, London (1982) 471-478.
17. J. Douglas, Jr., and T.F. Russell, Numerical methods for convection dominated problems based on combining the method of characteristics with finite element or finite difference procedures, SIAM J. Numer. Anal., 19 (1982) 871-885.

18. R.D. Richtmyer, A survey of difference methods for nonsteady fluid dynamics, NCAR Technical Note 63-2, Boulder, Colorado (1963).
19. S.Z. Burstein, Finite-difference calculations for hydrodynamic flows containing discontinuities, J. Comput. Physics, 2 (1967) 198-222.
20. J.T. Oden, T. Strouboulis and P. Devloo, Adaptive finite element methods for the analysis of inviscid compressible flow: Part I. Fast refinement/unrefinement and moving mesh methods for unstructured meshes, Comput. Meths. Appl. Mech. Engng., 59 (1986) 327-362.

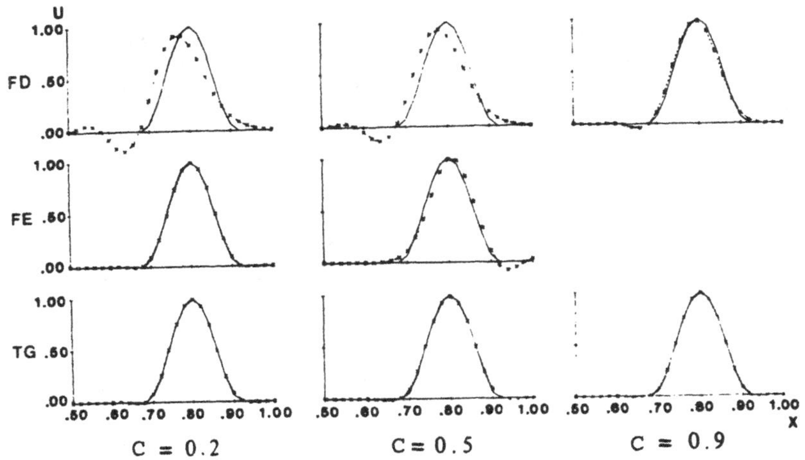

Fig. 1 - Propagation of a cosine profile. Comparison of the Lax-Wendroff (FD,FE) and Taylor-Galerkin (TG) schemes for several values of the Courant number.

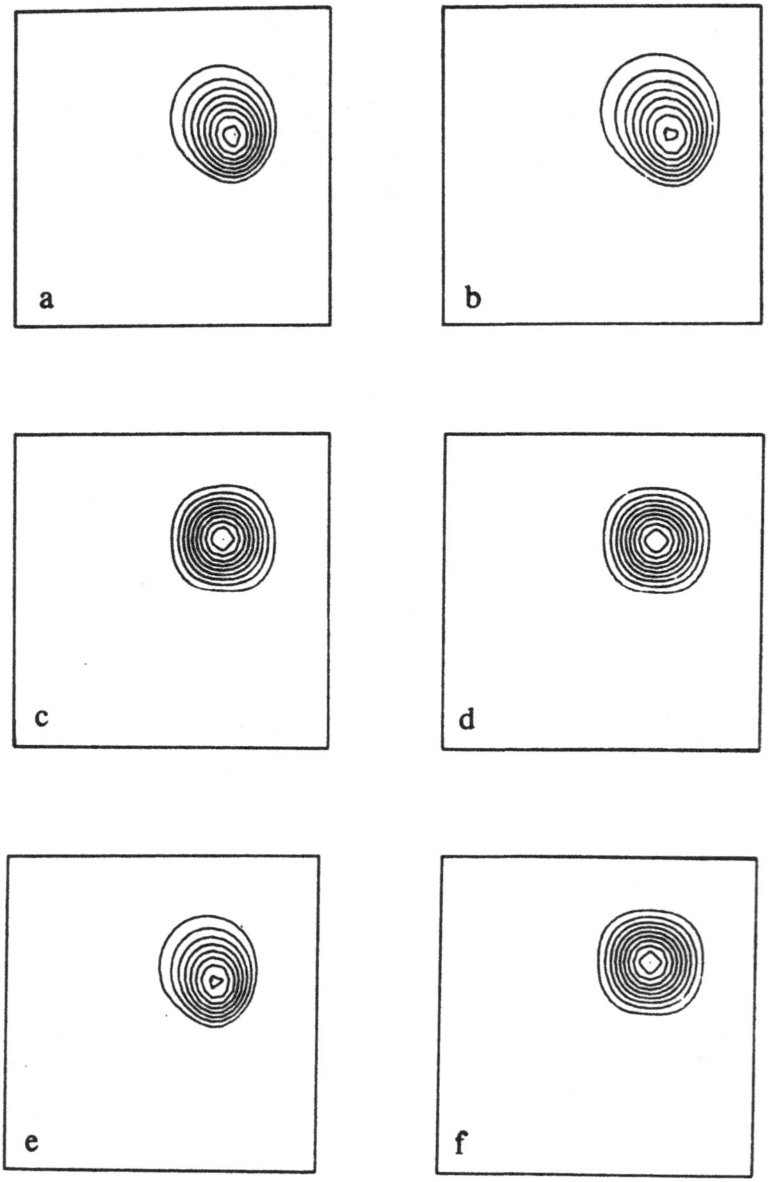

Fig. 2 - Advection of a cosine hill in a pure rotation velocity field: a) FD; b) FE (diagonal mass); c) FE (consistent mass); d) TG (one complete revolution in 200 time steps); e) FD; f) TG (one complete revolution in 120 time steps).

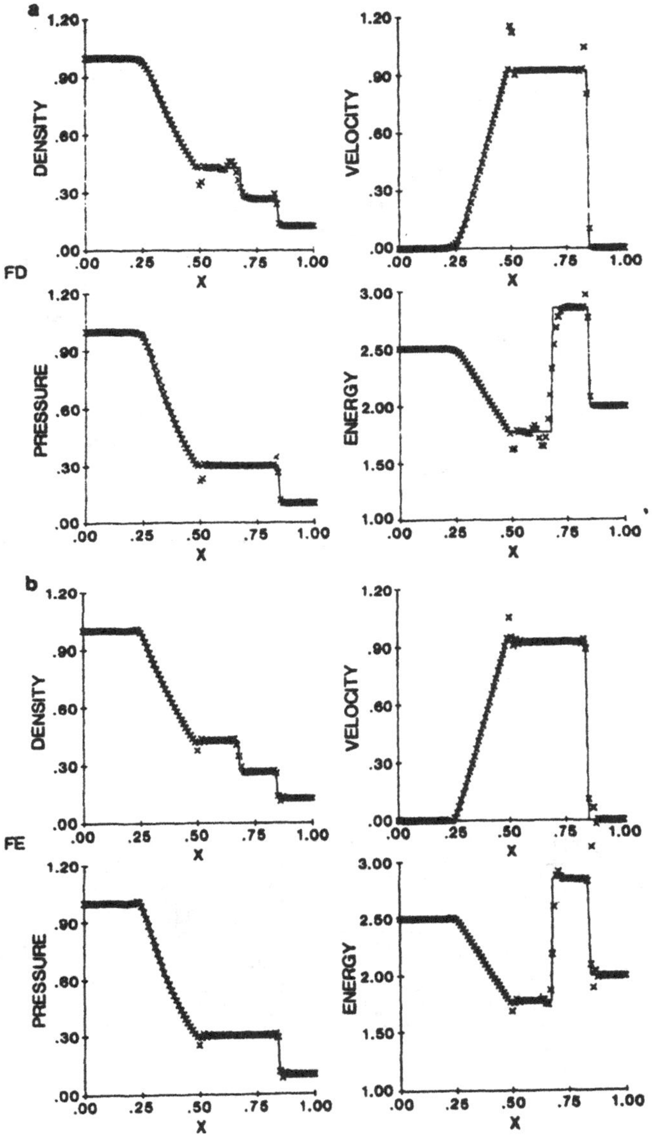

Fig. 3 - Numerical solutions of the Sod shock-tube problem by means of non-linear Lax-Wendroff schemes. Comparison of (a) two-step finite difference scheme (C = 0.9); (b) second-order finite element scheme (C = 0.45); and (c) third-order Taylor-Galerkin scheme (C = 0.9).

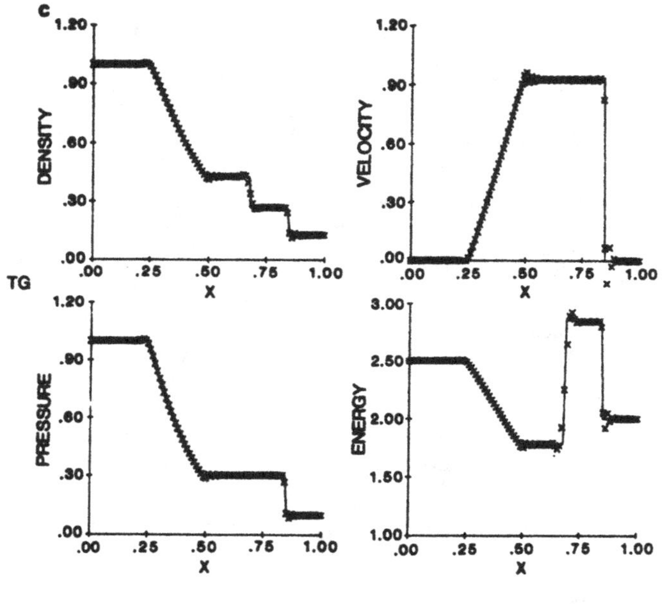

Fig. 3 - Continued.

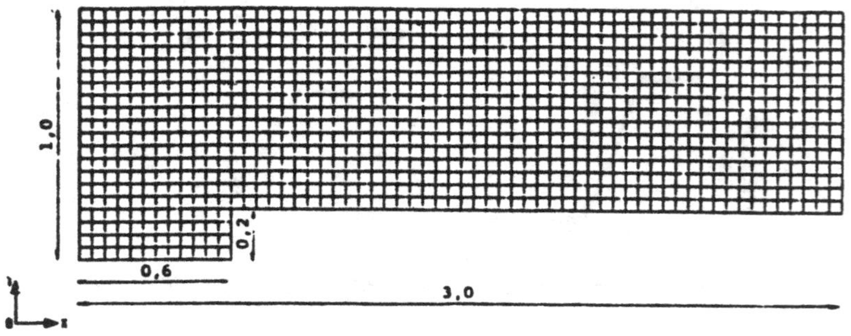

Fig. 4 - Transient supersonic flow (Mach 3) in a channel with a step. Finite element mesh of bilinear elements and computed pressure contours at several instants in time.

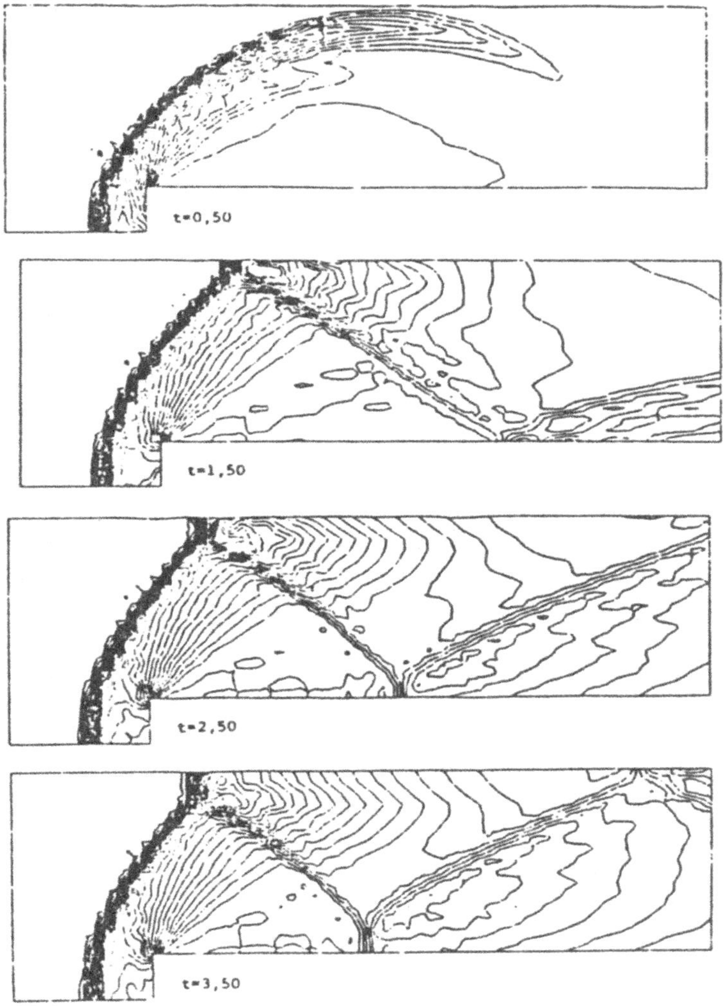

Fig. 4 - Continued.

CHAPTER 6

ACCELERATION PROCEDURES FOR THE NUMERICAL SIMULATION OF COMPRESSIBLE AND INCOMPRESSIBLE VISCOUS FLOWS

M. O. Bristeau
INRIA, Le Chesnay, France

R. Glowinski
University of Houston, Houston, Texas, USA

J. Periaux
AMD/BA, Saint Cloud, France

Abstract: Applying operator splitting methods to the numerical simulation of compressible or incompressible viscous flows leads to the solution of Stokes type linear problems and of nonlinear elliptic systems. Once discretized, these problems involve a large number of variables and therefore require efficient solution methods.

In this paper, we discuss the solution of the Stokes subproblems by iterative methods preconditioned by a well chosen boundary operator. The nonlinear subproblems (which are highly advective) are solved by an iterative method of *GMRES* type (cf. [1]). Numerical results corresponding to the simulation of flow around (and/or inside) air intakes and bodies illustrate the possibilities of the methods discussed here.

1. Introduction and Synopsis

The numerical simulation of two and three dimensional viscous flows modeled by the ***Navier-Stokes equations*** can be achieved via ***operator splitting methods***, like those described in [2] and [3]. The resulting subproblems are themselves fairly complicated requiring, therefore, efficient solution methods.

The main goal of this paper is to describe some novel iterative methods for the solution of the compressible and incompressible Navier-Stokes equations via operator splitting.

2. The Navier-Stokes equations for viscous flows.

We consider first the (important) particular case of flows around airfoils. Using the notation of Figure 2.1, ***isothermal, incompressible viscous*** flows are modeled by the following Navier-Stokes partial differential equations in the flow region Ω (we suppose that the ***density*** ρ is equal to 1):

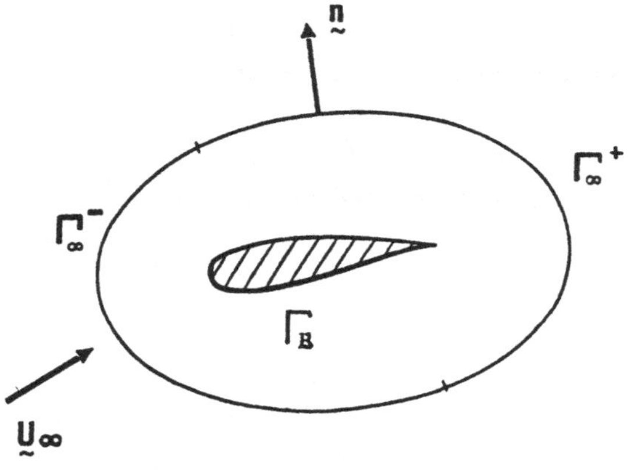

Figure 2.1

$$\frac{\partial u}{\partial t} - \nu\nabla^2 u + (u\cdot\nabla)\, u + \nabla p = 0 \text{ in } \Omega, \quad (2.1)$$

(2.2) $\nabla \cdot \mathbf{u} = 0$ *in* Ω (*incompressibility condition*),

with $\mathbf{u} = \{u_i\}_{i=1}^{N}$ (N = 2,3 in practice) the *flow velocity*, p the *pressure* and ν a *viscosity coefficient*, and

$$(\mathbf{u}\cdot\nabla)\mathbf{u} = \left\{\sum_{j=1}^{N} u_j \frac{\partial u_i}{\partial x_j}\right\}_{i=1}^{N}.$$

Boundary conditions have to be added to (2.1), (2.2). For simplicity, we shall only consider *Dirichlet conditions*, i.e.

(2.3) $\mathbf{u} = \mathbf{g}$ *on* Γ ($\Gamma = \partial\Omega$),

with $\int_\Gamma \mathbf{g}\cdot\mathbf{n}\, d\Gamma = 0$; $\mathbf{n}$: unit vector of the outward normal at Γ. The implementation of boundary conditions more complicated than (2.3) is discussed in, e.g., [2], [3]. An *initial condition* also has to be added to (2.1) - (2.3); we shall take here

(2.4) $\mathbf{u}(\mathbf{x}, 0) = \mathbf{u}_0(\mathbf{x})$ (*with* $\nabla\cdot\mathbf{u}_0 = 0$).

For *compressible viscous flows*, we shall use the following (*non conservative*) model:

(2.5) $$\frac{\partial \rho}{\partial t} + \nabla\cdot(\rho\,\mathbf{u}) = 0,$$

(2.6) $$\rho\frac{\partial \mathbf{u}}{\partial t} + \rho\,(\mathbf{u}\cdot\nabla)\,\mathbf{u} + \nabla p = \frac{1}{Re}[\nabla^2\mathbf{u} + \frac{1}{3}\nabla(\nabla\cdot\mathbf{u})],$$

(2.7) $$\rho\frac{\partial T}{\partial t} + \rho\,\mathbf{u}\cdot\nabla T + p\nabla\cdot\mathbf{u} = \frac{1}{Re}[\frac{\gamma}{Pr}\nabla^2 T + F(\nabla\mathbf{u})],$$

where the pressure satisfies the *perfect gas law*

(2.8) $$p = (\gamma - 1)\,\rho\,T,$$

where ρ, u, T are the *normalized density, velocity and temperature*, respectively, with (if N = 2)

(2.9) $$F(\nabla\mathbf{u}) = \frac{4}{3}\left[\left(\frac{\partial u_1}{\partial x_1}\right)^2 + \left(\frac{\partial u_2}{\partial x_2}\right)^2 - \frac{\partial u_1}{\partial x_1}\frac{\partial u_2}{\partial x_2}\right] + \left(\frac{\partial u_1}{\partial x_2} + \frac{\partial u_2}{\partial x_1}\right)^2.$$

In (2.6), (2.7), Re, Pr and γ are the *Reynolds Number*, the *Prandtl Number*, and the *ratio of the specific heats* ($\gamma = 1.4$ *in air*).

With Γ_∞ as in Figure 2.1, and

(2.10) $\Gamma_\infty^- = \{x | x \in \Gamma_\infty, u_\infty \cdot n_\infty(x) < 0\}$,

(2.11) $\Gamma_\infty^+ = \{x | x \in \Gamma_\infty, u_\infty \cdot n_\infty(x) \geq 0\}$,

we require on the *upstream boundary* Γ_∞^- the following conditions

(2.12) $u = u_\infty, \rho = 1, \quad T = \dfrac{1}{\gamma(\gamma\text{-}1)\, M_\infty^2}$;

on the *downstream boundary* Γ_∞^+ we require natural boundary conditions. Finally, on Γ_B we prescribe

(2.13) $u = 0$ *and* $T = T_\infty\,(1 + \frac{\gamma-1}{2} M_\infty^2)$.

In (2.10), (2.12), (2.13), u_∞, T_∞, M_∞ denote the velocity, the temperature and the Mach number for the corresponding *free flow* i.e., the one holding in the far field.

Finally, we shall also prescribe the following *initial conditions*

(2.14) $\rho(x, 0) = \rho_o(x)$, $u(x, 0) = u_o(x)$, $T(x, 0) = T_o(x)$.

In order to increase the commonality between the incompressible and compressible cases, it is convenient to introduce the new variable

(2.15) $\sigma = \ln \rho$.

With this new variable, the Navier Stokes equations become

(2.16) $\dfrac{\partial \sigma}{\partial t} + \nabla \cdot u + u \cdot \nabla \sigma = 0$,

$$(2.17)\qquad \frac{\partial u}{\partial t} + (u\cdot\nabla)u + (\gamma - 1)(T\,\nabla\sigma + \nabla\,T) = \frac{e^{-\sigma}}{Re}\left(\nabla^2 u + \frac{1}{3}\nabla(\nabla\cdot u)\right),$$

$$(2.18)\qquad \frac{\partial T}{\partial t} + u\cdot\nabla\,T + (\gamma - 1)\,T\,\nabla\cdot u = \frac{e^{-\sigma}}{Re}\left(\frac{\gamma}{Pr}\nabla^2 T + F(\nabla u)\right).$$

Equations (2.17), (2.18) can also be written as

$$(2.19)\qquad \frac{\partial u}{\partial t} - \mu\,\nabla^2 u + \beta\,\nabla\sigma - \psi(\sigma, u, T) = 0,$$

$$(2.20)\qquad \frac{\partial T}{\partial t} - \Pi\,\nabla^2\,T - \chi(\sigma, u, T) = 0,$$

respectively, with:

(a) δ an averaged value of the reciprocal of the density ($\delta = 1$ is a possible value),

(b) $\nu = 1/Re$, $\mu = \nu\delta$, $\Pi = \gamma\nu\delta/Pr$,

(c) $\beta = (\gamma\text{-}1)\,T_B = \frac{1}{\gamma}\left(\frac{\gamma-1}{2} + \frac{1}{M_\infty^2}\right)$,

(d) $$\psi(\sigma, u, T) = -(\gamma - 1)\,[\nabla\,T + (T - T_B)\,\nabla\sigma]$$

$$-(u\cdot\nabla)u + \nu[e^{-\sigma}(\nabla^2 u + \frac{1}{3}\nabla(\nabla\cdot u)) - \delta\,\nabla^2 u],$$

(e) $\chi(\sigma, u, T) = -(\gamma - 1)\,T\nabla\cdot u - u\cdot\nabla T + \frac{\gamma\nu}{Pr}(e^{-\sigma} - \delta)\,\nabla^2\,T + \nu\,e^{-\sigma}F(\nabla u)$.

3. Time Discretization by Operator Splitting.

Using time discretization by operator splitting, we can decouple (as shown in e.g. [2], [3]) the various operators occuring in the above mathematical models. In the *incompressible* case, we should obtain the following scheme (of *Peaceman - Rachford* type) where Δt (> 0) is the time discretization step:

$$(3.1)\qquad u^o = u_o;$$

then for $n \geq 0$, *assuming that* u^n *is known, compute* $\{u^{n+\frac{1}{2}}, p^{n+\frac{1}{2}}\}$, *and* u^{n+1} *by*

(3.2)
$$\begin{cases} \dfrac{u^{n+\frac{1}{2}} - u^n}{\Delta t/2} - \dfrac{\nu}{2} \nabla^2 u^{n+\frac{1}{2}} + \nabla p^{n+\frac{1}{2}} = \dfrac{\nu}{2} \nabla^2 u^n - (u^n \cdot \nabla)\, u^n \quad \textit{in } \Omega, \\ \nabla \cdot u^{n+\frac{1}{2}} = 0 \quad \textit{in } \Omega, \\ u^{n+\frac{1}{2}} = g^{n+\frac{1}{2}} \quad \textit{on } \Gamma, \end{cases}$$

(3.3)
$$\begin{cases} \dfrac{u^{n+1} - u^{n+\frac{1}{2}}}{\Delta t/2} - \dfrac{\nu}{2} \nabla^2 u^{n+1} + (u^{n+1} \cdot \nabla)\, u^{n+1} = \dfrac{\nu}{2} \nabla^2 u^{n+\frac{1}{2}} - \nabla p^{n+\frac{1}{2}} \quad \textit{in } \Omega, \\ u^{n+1} = g^{n+1} \quad \textit{on } \Gamma, \end{cases}$$

respectively.

We observe that the subproblems (3.2) are *linear* and quite close to steady Stokes problems; concerning the advection-diffusion subproblems (3.3) they are *nonlinear* elliptic systems.

Remark 3.1 A useful (and cheaper) *linear* variant of (3.3) is obtained by substituting the linear term $(u^{n+\frac{1}{2}} \cdot \nabla)\, u^{n+1}$ to $(u^{n+1} \cdot \nabla)\, u^{n+1}$ in (3.3). □

More sophisticated operator splitting methods for the incompressible Navier-Stokes equations are discussed in e.g. [2], [3]; in fact their *compressible* variant, applied to (2.16) - (2.20), is given, for $\theta \in (0, 1/2)$, by

(3.4) $\sigma^0 = \sigma_0 = \ln \rho_0$, $u^0 = u_0$, $T^0 = T_0$;

then for $n \geq 0$, *assuming that* σ^n, u^n, T^n *are known we solve successively*

$$(3.5)\quad \begin{cases} \dfrac{\sigma^{n+\theta}-\sigma^{n}}{\theta\Delta t} + \nabla\cdot u^{n+\theta} = -u^{n}\cdot\nabla\sigma^{n}, \\[2ex] \dfrac{u^{n+\theta}-u^{n}}{\theta\Delta t} - a\mu\,\nabla^2 u^{n+\theta} + \beta\nabla\sigma^{n+\theta} = \psi(\sigma^{n}, u^{n}, T^{n}) + b\,\mu\nabla^2 u^{n}, \\[2ex] \dfrac{T^{n+\theta}-T^{n}}{\theta\Delta t} - a\,\Pi\,\nabla^2 T^{n+\theta} = \chi(\sigma^{n}, u^{n}, T^{n}) + b\,\Pi\nabla^2 T^{n}, \end{cases}$$

$$(3.6)\quad \begin{cases} \dfrac{\sigma^{n+1-\theta}-\sigma^{n+\theta}}{(1-2\theta)\Delta t} + u^{n+1-\theta}\cdot\nabla\sigma^{n+1-\theta} = -\nabla\cdot u^{n+\theta}, \\[2ex] \dfrac{u^{n+1-\theta}-u^{n+\theta}}{(1-2\theta)\Delta t} - b\mu\nabla^2 u^{n+1-\theta} - \psi(\sigma^{n+1-\theta}, u^{n+1-\theta}, T^{n+1-\theta}) = a\mu\nabla^2 u^{n+\theta} - \beta\nabla\sigma^{n+\theta}, \\[2ex] \dfrac{T^{n+1-\theta}-T^{n+\theta}}{(1-2\theta)\,\Delta t} - b\Pi\,\nabla^2 T^{n+1-\theta} - \chi(\sigma^{n+1-\theta}, u^{n+1-\theta}, T^{n+1-\theta}) = a\,\Pi\,\nabla^2 T^{n+\theta}, \end{cases}$$

$$(3.7)\quad \begin{cases} \dfrac{\sigma^{n+1}-\sigma^{n+1-\theta}}{\theta\Delta t} + \nabla\cdot u^{n+1} = -u^{n+1-\theta}\cdot\nabla\sigma^{n+1-\theta}, \\[2ex] \dfrac{u^{n+1}-u^{n+1-\theta}}{\theta\Delta t} - a\,\mu\nabla^2 u^{n+1} + \beta\,\nabla\sigma^{n+1} = \psi(\sigma^{n+1-\theta}, u^{n+1-\theta}, T^{n+1-\theta}) + \\ b\mu\,\nabla^2 u^{n+1-\theta}, \\[2ex] \dfrac{T^{n+1}-T^{n+1-\theta}}{\theta\Delta t} - a\Pi\,\nabla^2 T^{n+1} = \chi(\sigma^{n+1-\theta}, u^{n+1-\theta}, T^{n+1-\theta}) + b\Pi\,\nabla^2 T^{n+1-\theta}, \end{cases}$$

with $0 < a, b < 1$, $a+b = 1$; a natural choice for a and b is

(3.8) $a = (1-2\theta)/(1-\theta), \; b = \theta/(1-\theta)$,

since with such a choice, the linear parts of the elliptic operators occuring in (3.5) - (3.7) are proportional. Of course, the above equations have to be completed by appropriate boundary conditions.

The above scheme, is a θ-scheme very close to those used for *incompressible* flows, and described in [3].

4. Accelerating the Convergence for the Stokes Subproblems

4.1. The Incompressible Case.

Subproblems (3.2) are particular case of the *generalized Stokes problem*

$$(4.1) \quad \begin{cases} \alpha u - \nu\nabla^2 u + \nabla p = f \text{ in } \Omega, \\ \nabla\cdot u = 0 \text{ in } \Omega, \\ u = g \text{ on } \Gamma. \end{cases}$$

Problem (4.1) - after an appropriate discretization - leads to linear systems which can be solved by classical *direct* methods such as Gaussian elimination; however, for large size problems, which is almost always the case, even for two-dimensional flow, the lack of symmetry of the corresponding matrix and the fact that it is indefinite make its solution fairly costly.

Therefore, we shall use iterative methods taking advantage of the special structure of (4.1); such methods are described in e.g. [2], [3] (see also [16]). In this paper, we shall describe a *conjugate gradient* method to solve (4.1), in which the *master unknown* is the *trace* of the pressure on the boundary Γ. This method is based on the Helmholtz decomposition of the Stokes problem (4.1), and uses - as *preconditioner* - a boundary operator quite easy to implement.

Construction of the preconditioning operator: Let λ be the trace of the pressure p on Γ; it is classical (cf. e.g. [2], [4]) and easy to check, that λ is solution of the following equation

(4.2) $\quad A\lambda = \beta,$

where A is the *boundary operator* defined as follows:

(4.3)$_1$ $\quad \nabla^2 p_\mu = 0$ *in* Ω, $p_\mu = \mu$ *on* Γ,

(4.3)$_2$ $\quad \alpha u_\mu - \nu\nabla^2 u_\mu = -\nabla p_\mu$ *in* Ω, $u_\mu = 0$ *on* Γ,

(4.3)$_3$ $\quad -\nabla^2 \psi_\mu = \nabla\cdot u_\mu$ *in* Ω, $\psi_\mu = 0$ *on* Γ,

(4.3)$_4$ $\quad A\mu = -\frac{\partial\psi_\mu}{\partial n}\,|_\Gamma;$

the right hand side β, in (4.2), is defined in turn by

(4.4)$_1$ $\quad \nabla^2 p_0 = \nabla\cdot f$ *in* Ω, $p_0 = 0$ *on* Γ ,

(4.4)$_2$ $\quad \alpha u_0 - \nu\nabla^2 u_0 = f - \nabla p_0$ *in* Ω, $u_0 = g$ *on* Γ ,

(4.4)$_3$ $\quad -\nabla^2 \psi_0 = \nabla\cdot u_0$ *in* Ω, $\psi_0 = 0$ *on* Γ,

(4.4)$_4$ $\quad \beta = \frac{\partial\psi_0}{\partial n}\,|_\Gamma$.

Operator A is not known explicitely, in general; it can be proved, however, that A is symmetric and *positive semi-definite* (cf. [2], [4] for these properties).

In order to solve (4.1) via (4.2), we can use a *conjugate gradient algorithm, preconditioned* by the operator S , defined as follows:

(4.5) $\quad S = B^{-1},$

where

(4.6) $\quad B\mu = \frac{\alpha}{\nu}\,\varphi_\mu|_\Gamma + 4\sqrt{\frac{\alpha}{\nu}}\,\mu + 4\,\frac{\partial\theta_\mu}{\partial n}\,|_\Gamma,$

with

$$(4.7) \qquad \nabla^2\varphi_\mu = 0 \textit{ in } \Omega,\ \frac{\partial\varphi_\mu}{\partial n} = \mu \textit{ on } \Gamma;\ \int_\Gamma \varphi_\mu \, d\Gamma = 0,$$

$$(4.8) \qquad \nabla^2\theta_\mu = 0 \textit{ in } \Omega,\ \theta_\mu = \mu \textit{ on } \Gamma.$$

A *Fourier Analysis* done when Ω is the half-space (cf. [5], [6]) shows that operator S defined by (4.5) - (4.8) is *quasi-optimal* as a preconditioner. The corresponding conjugate gradient algorithm to solve (4.1), via (4.2), is given by

Step 0: Initialization

$$(4.9) \qquad \lambda^0 \textit{ is given;}$$

we solve then

$$(4.10) \qquad \nabla p^0 = \nabla\cdot \mathbf{f} \text{ in } \Omega,\ p^0 = \lambda^0 \textit{ on } \Gamma,$$

$$(4.11) \qquad \alpha \mathbf{u}^0 - \nu\nabla^2\mathbf{u}^0 = \mathbf{f} - \nabla p^0 \textit{ in } \Omega,\ \mathbf{u}^0 = \mathbf{g} \textit{ on } \Gamma,$$

$$(4.12) \qquad -\nabla^2\psi^0 = \nabla\cdot \mathbf{u}^0 \textit{ in } \Omega,\ \psi^0 = 0 \textit{ on } \Gamma,$$

and we define

$$(4.13) \qquad r^0 = -\frac{\partial\psi^0}{\partial n}\Big|_\Gamma .$$

Preconditioning is obtained by solving

$$(4.14)_1 \qquad \nabla^2\varphi^0 = 0 \textit{ in } \Omega,\ \frac{\partial\varphi^0}{\partial n} = r^0 \textit{ on } \Gamma;\ \int_\Gamma \varphi^0 d\Gamma = 0,$$

$$(4.14)_2 \qquad \nabla^2\theta^0 = 0 \textit{ in } \Omega,\ \theta^0 = r^0 \textit{ on } \Gamma,$$

and by setting

(4.15) $$g^0 = \frac{\alpha}{\nu}\varphi^0|_\Gamma + 4\sqrt{\frac{\alpha}{\nu}}\, r^0 + 4\,\frac{\partial\theta^0}{\partial n}\,|_\Gamma,$$

(4.16) $$w^0 = g^0 .$$ □

Then for $n \geq 0$ *with* $\lambda^n, p^n, u^n, \psi^n, r^n, g^n, w^n$ *known compute* $\lambda^{n+1}, p^{n+1}, u^{n+1}, \psi^{n+1}, r^{n+1}, g^{n+1}, w^{n+1}$ *as follows:*

Step 1: Descent

(4.17) $$\nabla^2 \bar{p}^n = 0 \textit{ in } \Omega , \quad \bar{p}^n = w^n \textit{ on } \Gamma,$$

(4.18) $$\alpha\, \bar{u}^n - \nu\, \nabla^2 \bar{u}^n = - \nabla \bar{p}^n \textit{ in } \Omega , \quad \bar{u}^n = 0 \text{ on } \Gamma,$$

(4.19) $$-\nabla^2 \bar{\psi}^n = \nabla\cdot\bar{u}^n \textit{ in } \Omega , \quad \bar{\psi}^n = 0 \textit{ on } \Gamma,$$

(4.20) $$\rho_n = (-1)\,\frac{\displaystyle\int_\Gamma r^n g^n \, d\Gamma}{\displaystyle\int_\Gamma \frac{\partial\bar{\psi}^n}{\partial n}\, w^n \, d\Gamma} .$$

Next we define $\lambda^{n+1}, p^{n+1}, u^{n+1}, \psi^{n+1}$ *and* r^{n+1} *by*

(4.21) $$\lambda^{n+1} = \lambda^n - \rho_n\, w^n ,$$

(4.22) $$p^{n+1} = p^n - \rho_n\, \bar{p}^n ,$$

(4.23) $$u^{n+1} = u^n - \rho_n\, \bar{u}^n ,$$

(4.24) $$\psi^{n+1} = \psi^n - \rho_n\, \bar{\psi}^n ,$$

(4.25) $$r^{n+1} = r^n + \rho_n\, \frac{\partial\bar{\psi}^n}{\partial n} .$$

Step 2: Construction of the new descent direction

We solve

$$\text{(4.26)}\quad \nabla^2\bar{\varphi}^{n+1} = 0 \text{ in } \Omega,\ \frac{\partial\bar{\varphi}^{n+1}}{\partial n} = r^{n+1} \text{ on } \Gamma;\ \int_\Gamma \bar{\varphi}^{n+1}\, d\Gamma = 0,$$

$$\text{(4.27)}\quad \nabla^2\bar{\theta}^{n+1} = 0 \text{ in } \Omega,\ \bar{\theta}^{n+1} = r^{n+1} \text{ on } \Gamma,$$

and we define g^{n+1} *by*

$$\text{(4.28)}\quad g^{n+1} = \frac{\alpha}{\nu}\,\bar{\varphi}^{n+1}|_\Gamma + 4\sqrt{\frac{\alpha}{\nu}}\, r^{n+1} + 4\,\frac{\partial\bar{\theta}^{n+1}}{\partial n}\,|_\Gamma \,.$$

We compute then

$$\text{(4.29)}\quad \gamma_n = \frac{\displaystyle\int_\Gamma r^{n+1} g^{n+1}\, d\Gamma}{\displaystyle\int_\Gamma r^n g^n\, d\Gamma}\,,$$

$$\text{(4.30)}\quad w^{n+1} = g^{n+1} + \gamma_n\, w^n\,. \quad \square$$

Do n = n+1 *and go to* (4.17).

4.2 The Compressible Case.

Subproblems (3.5) and (3.7) are particular cases of the following class of systems of partial differential equations

$$\text{(4.31)}\quad \alpha\sigma + \nabla\cdot u = g \text{ in } \Omega,$$

$$\text{(4.32)}\quad \alpha u - a\mu\nabla^2 u + \beta\,\nabla\sigma = f \text{ in } \Omega,$$

$$\text{(4.33)}\quad \alpha T - a\,\Pi\,\nabla^2 T = h \text{ in } \Omega,$$

for the following boundary conditions:

$$(4.34) \qquad u = 0 \ on \ \Gamma_B, \ u = u_\infty \ on \ \Gamma_\infty^-, \ a\mu \frac{\partial u}{\partial n} - \beta\sigma n = s \ on \ \Gamma_\infty^+ ,$$

$$(4.35) \qquad T = T_B \ on \ \Gamma_B, \ T = T_\infty \ on \ \Gamma_\infty^-, \ \frac{\partial T}{\partial n} = 0 \ on \ \Gamma_\infty^+ ;$$

in (4.31) - (4.33), α is positive parameter.

The solution of the elliptic problem (4.33), (4.35) is fairly classical and will not be discussed in this paper; concerning now (4.31), (4.32), (4.34) we observe that this elliptic system is indeed a generalization of the Stokes problem (4.1); actually similar systems occur in the solution of the *shallow water equations* (cf. [7] and the references therein) . The methodology developed in section 4.1 for incompressible flow can be generalized here. We shall therefore consider the solution of problem (4.31), (4.32), (4.34) by methods either *direct* or *iterative*, founded on the (Helmholtz) decomposition below; the goal in this approach is to reduce the solution of the above problem to that of a finite number of simple elliptic problems, coupled to the the solution of a boundary integral equation on $\Gamma(= \Gamma_B \cup \Gamma_\infty^+ \cup \Gamma_\infty^-)$. Taking the divergence of both sides in (4.32), we obtain

$$(4.36) \qquad \alpha\nabla\cdot u - a\,\mu\Delta\,(\nabla\cdot u) + \beta\,\Delta\sigma = \nabla\cdot f ;$$

on the other hand, it follows from (4.31) that

$$(4.37) \qquad \nabla\cdot u = g - \alpha\sigma .$$

Combining now (4.36), (4.37) we obtain

$$(4.38) \qquad \alpha^2\sigma - (\beta + \alpha a\mu)\,\Delta\sigma = \alpha g - \nabla\cdot f - a\mu\,\Delta g .$$

To the elliptic equation (4.38), we associate the Dirichlet boundary condition

$$(4.39) \qquad \sigma = \lambda \ on \ \Gamma .$$

If λ is known, we obtain σ from (4.38), then u from (4.32), (4.34). In order to find λ such that (4.31) holds, it is convenient to introduce ψ solution of

$$(4.40) \qquad \alpha^2\psi - \zeta\Delta\psi = \alpha\sigma + \nabla\cdot \mathbf{u} - g \text{ in } \Omega ,$$

$$(4.41) \qquad \psi = 0 \text{ on } \Gamma ,$$

with $\zeta = \beta + \alpha a\mu$. We can easily show that ψ satisfies the following *biharmonic equation*

$$(4.42) \qquad \alpha^2\psi - (\zeta + \alpha a\mu)\Delta\psi + \alpha a\mu\zeta\, \Delta^2\psi = 0 ;$$

if we can prove that $\psi \equiv 0$, then (4.31) will be satisfied. A sufficient condition, is clearly to chose λ such that

$$(4.43) \qquad \frac{\partial\psi}{\partial n} = 0 \text{ on } \Gamma.$$

Let's show the feasibility of this approach; to do so we introduce a *linear operator* A, defined as follows:

The function λ being given over Γ, we solve the following elliptic problems:

$$(4.44) \qquad \begin{cases} \alpha^2\sigma_\lambda - \zeta\Delta\sigma_\lambda = 0 \text{ in } \Omega , \\ \sigma_\lambda = \lambda \text{ on } \Gamma , \end{cases}$$

$$(4.45) \qquad \begin{cases} \alpha \mathbf{u}_\lambda - a\mu\, \Delta \mathbf{u}_\lambda = -\beta\nabla\sigma_\lambda \text{ in } \Omega , \\ \mathbf{u}_\lambda = 0 \text{ on } \Gamma_B \cup \Gamma_\infty^- , \\ a\mu \dfrac{\partial \mathbf{u}_\lambda}{\partial n} - \beta \mathbf{n}\sigma_\lambda = 0 \text{ on } \Gamma_\infty^+ . \end{cases}$$

$$(4.46)\quad \begin{cases} \alpha^2\psi_\lambda - \zeta\Delta\psi_\lambda = \alpha\sigma_\lambda + \nabla\cdot \mathbf{u}_\lambda \ \textit{in}\ \Omega, \\ \psi_\lambda = 0 \ \textit{on}\ \Gamma, \end{cases}$$

and we define A by

$$(4.47)\qquad A\lambda = -\frac{\partial\psi_\lambda}{\partial n}\Big|_\Gamma .$$

Operator A is *symmetric* and *positive definite*; indeed, it follows from (4.44) - (4.46), and from the Green's formula that

$$(4.49)\qquad <A\lambda_1, \lambda_2> = \frac{\alpha}{\zeta}\int_\Omega \sigma_1\sigma_2 dx + \frac{1}{\beta\zeta}\int_\Omega [\alpha \mathbf{u}_1 : \mathbf{u}_2 + a\mu\, \nabla \mathbf{u}_1 \cdot \nabla \mathbf{u}_2]\, dx,$$

where σ_1, $\mathbf{u}_1$(resp. σ_2, $\mathbf{u}_2$) are the solutions of (4.44), (4.45) associated to λ_1 (resp. λ_2). From (4.49), operator A is clearly *symmetric* ; in the same fashion, combining (4.49) and (4.44), (4.45), one easily shows that A is *positive definite*.

Let's apply now the above results to the solution of problem (4.31), (4.32), (4.33); to do so we introduce σ_o, $\mathbf{u}_o$ and ψ_o , solutions of

$$(4.50)\quad \begin{cases} \alpha^2\sigma_o - \zeta\Delta\sigma_o = \alpha g - \nabla\cdot \mathbf{f} - a\mu\Delta g \ \textit{in}\ \Omega, \\ \sigma_o = 0 \ \textit{on}\ \Gamma, \end{cases}$$

$$(4.51)\quad \begin{cases} \alpha \mathbf{u}_o - a\mu\, \Delta \mathbf{u}_o = \mathbf{f} - \beta\nabla\sigma_o \ \textit{in}\ \Omega, \\ \mathbf{u}_o = 0 \ \textit{on}\ \Gamma_B,\ \mathbf{u}_o = \mathbf{u}_\infty \ \textit{on}\ \Gamma_\infty^-, \\ a\mu\dfrac{\partial \mathbf{u}_o}{\partial n} = \mathbf{s} \ \textit{on}\ \Gamma_\infty^+, \end{cases}$$

$$(4.52)\quad \begin{cases} \alpha^2\psi_0 - \zeta\,\Delta\psi_0 = \alpha\,\sigma_0 + \nabla\cdot u_0 - g \text{ in } \Omega, \\ \\ \psi_0 = 0 \text{ on } \Gamma. \end{cases}$$

By substraction and comparison between (4.31), (4.32), (4.34) and (4.50) - (4.52), we can easily show that the trace λ of σ on Γ satisfies the (pseudo-differential) equation

$$(4.53)\qquad A\lambda = \frac{\partial\psi_0}{\partial n}\Big|_\Gamma .$$

A first possibility to solve (4.53) is to use a *quasi-direct* method in the sense of [2, Chapter 7] and [4] ; this approach requires the construction of some finite dimensional approximation of A. Since the above construction is quite costly for three dimensional flow problems, it is worth trying to generalize the methodology of Section 4.1 (incompressible case) in order to include the present compressible case. To achieve such a goal we can still use Fourier Analysis to solve exactly (4.31), (4.32), (4.34) when Ω is an half space with $\Gamma_B = \Gamma$. Similarly, the conjugate gradient algorithm (4.9) - (4.30) can be generalized to the above problem ; then, we shall use as preconditioner, the operator S defined by

$$(4.54)\qquad S^{-1} = B,$$

$$(4.55)\qquad Bz = c\,\phi_z\,|_\Gamma + 2z + \frac{1}{c}\frac{\partial\theta_z}{\partial n}\Big|_\Gamma$$

with

$$(4.56)\qquad \alpha^2\phi_z - \zeta\nabla^2\phi_z = 0 \text{ in } \Omega,\ \frac{\partial\phi_z}{\partial n} = z \text{ on } \Gamma,$$

$$(4.57)\qquad \alpha^2\theta_z - \zeta\nabla^2\theta_z = 0 \text{ in } \Omega,\ \theta_z = z \text{ on } \Gamma,$$

$$(4.58)\qquad c = \frac{\alpha}{\sqrt{\zeta}}\ \{\tfrac{1}{2}[(1+\frac{\zeta}{\alpha a\mu})\,(1+\sqrt{\frac{\alpha a\mu}{\zeta}}\,)]^{1/2} - 1\},$$

where

$$(4.59)\qquad \zeta = \beta + \alpha a\mu .$$

Parameter c appears naturally from the Fourier Analysis of operator A (cf. Appendix 1 and [8]). See also [19] for the description of a conjugate gradient implementation of the above operator B, applied to the solution of problem (4.31), (4.32).

5. Solution of the nonlinear problems by GMRES type algorithms.

5.1 Generalities

Among the various numerical methods which can be used for solving nonlinear problems of large dimension, let's mention nonlinear least squares methods, since these methods - coupled to conjugate gradient algorithms - have been successfully applied to the solution of complicated problems arising from fluid mechanics (cf. e.g. [2], [3], [9] for these applications).

One of the major drawbacks of the above methods is that they require an accurate knowledge of the gradient of the cost function; for some problems this knowledge is very costly in itself (for example, this seems to be the case for the compressible Navier-Stokes equations). Recently, several investigators have introduced variants of the above methods which do not require the exact knowledge of the gradient. Among these methods, GMRES (cf. [1]) has shown interesting possibilities for nonlinear problems (cf. [10], [11] for the theory and some applications of GMRES in Fluid Dynamics). In the next section we shall use the framework of abstract Hilbert spaces to describe a generalization of GMRES whose finite dimensional variants can be seen as preconditioned variations of the original algorithm. In fact, these methods have been already applied to the compressible Euler equations (cf. e.g. [11], [12]).

5.2 The GMRES algorithm for nonlinear problems.

Applying operator splitting methods to the Navier-Stokes equations leads to the solution of systems of nonlinear partial differential equations. Using appropriate finite difference, finite element or spectral methods the above systems are reduced to nonlinear systems in finite dimension. The continuous problems and the associate discrete ones are all relevant to the following framework:

Let's consider an Hilbert space V, whose scalar product and corresponding norm are denoted by $(.,.)$ and $\| \cdot \|$, respectively. We denote by V' the dual space of V, by $\langle .,. \rangle$ the duality pairing between V' and V, and finally, by S the duality isomorphism between V and V', i.e. the isomorphism from V onto V' satisfying:

$$\langle Sv, w \rangle = (v, w), \ \forall v, w \in V,$$

$$\langle Sv, w \rangle = \langle Sw, v \rangle, \ \forall v, w \in V,$$

$$<f, S^{-1}g> = <f, g>_* \ \forall f, g \in V'.$$

With F a (possibly nonlinear) operator from V to V' we consider the following problem

(5.1) $$F(u) = 0 .$$

Description of a GMRES algorithm for the solution of (5.1) :

(5.2) $u^o \in V$ *is given;*

Then, for $n \geq 0$, u^n *being known we obtain* u^{n+1} *as follows:*

(5.3) $$r_1^n = S^{-1}F(u^n) ,$$

(5.4) $$w_1^n = r_1^n / \| r_1^n \| .$$

Then, for $j = 2, \ldots, k$, *we compute* r_j^n *and* w_j^n *by*

(5.5) $$r_j^n = S^{-1}DF(u^n; w_{j-1}^n) - \sum_{i=1}^{j-1} b_{ij-1} w_i^n ,$$

(5.6) $$w_j^n = r_j^n \ / \ \| r_j^n \| ;$$

in (5.5) , $DF(u^n; w)$ *is defined by either*

$(5.7)_1$ $$DF(u^n; w) = F'(u^n) . w$$

(*where* $F'(u^n)$ *is the derivative of* F *at* u^n) , *or, if the calculation of* $F'(u^n)$ *is too costly, by*

$(5.7)_2$ $$DF(u^n; w) = \frac{F(u^n + \varepsilon\, w) - F(u^n)}{\varepsilon}$$

with $\varepsilon (> 0)$ *sufficiently small; we define then* b_{il}^n *by*

(5.8) $$b_{il}^n = <DF(u^n; w_l^n), \ w_i^n> \ (= (S^{-1}DF(u^n; w_l^n) , w_i^n)) .$$

Then we solve

$$(5.9)\quad \begin{cases} \textit{Find } a^n = \{a_j^n\}_{j=1}^k \varepsilon\, R^k, \textit{ such that} \\ \forall\, c = \{c_j\}_{j=1}^k \;\varepsilon R^k, \textit{ one has} \\ \| F(u^n + \sum_{j=1}^k a_j^n\, w_j^n)\|_* \leq \| F\,(u^n + \sum_{j=1}^k c_j\, w_j^n)\|_* \end{cases}$$

and obtain u^{n+1} *by*

$$(5.10)\qquad u^{n+1} = u^n + \sum_{j=1}^k a_j^n\, w_j^n \,. \;\square$$

Do $n = n+1$ *and go to* **(5.3)**.

In algorithm (5.2) - (5.10), k is the dimension of the so called *Krylov* space.

Remark 5.1 : We can easily show that

$$(5.11)\qquad (w_j^n,\, w_l^n) = 0,\ \forall 1 \leq l,j \leq k,\, j \neq l\,.$$

Remark 5.2 : To compute $DF(u^n;\, w)$, instead of $(5.7)_2$, we can use the following second order accurate approximation of $F'(u^u)$. w

$$(5.12)\qquad DF(u^n;\, w) = \frac{F(u^n + \varepsilon\, w) - F(u^n - \varepsilon\, w)}{2\varepsilon}\,;$$

the choice of ε in $(5.7)_2$ and (5.12) can be done automatically by checking the variation of DF $(u^n,\, w)$ (cf. e.g. [13]) .

Remark 5.3 : The $\|\, f\,\|_*$ norm in (5.9) satisfies the following relations

$$(5.13)\qquad \|\, f\,\|_* = \|\, S^{-1} f\|\,,\ \forall f\ \varepsilon\ V'\,,$$

$$(5.14)\qquad \|\, f\,\|_* = \langle f,\ S^{-1} f\rangle^{1/2}\,,\ \forall f\ \varepsilon\ V'\,;$$

in practice, one uses (5.13) to evaluate the various $\|\cdot\|_*$ norms occuring in (5.9).

Remark 5.4 : To compute a^n from the solution of (5.9), it suffices (in general) to approximate (in the neighborhood of $c = 0$) the functional

$$(5.15)_1 \quad c \to \| F(u^n + \sum_{j=1}^{k} c_j w_j^n) \|_*$$

by the *quadratic* one defined by

$$(5.15)_2 \quad c \to \| F(u^n) + \sum_{j=1}^{k} c_j DF(u^n; w_j^n) \|_*^2.$$

Minimizing $(5.15)_2$ is equivalent to solving a linear system with a symmetric and positive definite $k \times k$ matrix.

5.3 Application to the numerical simulation of viscous flow.

For simplicity, we consider the *incompressible* case only. It follows from Section 3 that operator splitting methods lead to the solution of nonlinear elliptic problems such as

$$(5.16) \quad \begin{cases} \alpha u - \nu \nabla^2 u + (u \cdot \nabla) u = f \textit{ in } \Omega, \\ u = g \textit{ on } \Gamma. \end{cases}$$

A variational formulation of (5.16) is given by

$$(5.17) \quad \begin{cases} u \in V_g \; (= \{v \mid v \in H^1(\Omega)^N, \; v = g \textit{ on } \Gamma\}), \\ \alpha \int_\Omega u \cdot v \, dx + \nu \int_\Omega \nabla u \cdot \nabla v \, dx + \int_\Omega (u \cdot \nabla) u \cdot v \, dx = \int_\Omega f \cdot v \, dx, \; \forall v \in V_o, \end{cases}$$

where $V_o = H_0^1(\Omega)^N$.

Actually, problem (5.16), (5.17) is a particular case of (5.1) ; to solve (5.16), (5.17) by the GMRES algorithm, we consider V_g as a subspace of $H^1(\Omega)^N$, this last space being equipped with the following scalar product

$$\{v, w\} \to \alpha \int_\Omega v \cdot w \, dx + \nu \int_\Omega \nabla v \cdot \nabla w \; dx,$$

and the associated norm

$$v \to \{ \alpha \int_\Omega | v |^2 dx + \nu \int_\Omega | \nabla v |^2 dx \}^{1/2} .$$

Once this choice has been done, applying algorithm (5.2) - (5.10) is fairly easy and we obtain then the following algorithm

Step 0: Initialization

(5.18) $\quad u^\circ \in V_g$ *is given ;*

then for $n \geq 0$, u^n *being known, we obtain* u^{n+1} *as follows*

Step 1: Construction of the Krylov space

Solve

$$(5.19) \quad \begin{cases} \alpha \int_\Omega r_1^n \cdot v \, dx + \nu \int_\Omega \nabla r_1^n \cdot \nabla v \, dx = \alpha \int_\Omega u^n \cdot v \, dx + \nu \int_\Omega \nabla u^n \cdot \nabla v \, dx \\ + \int_\Omega (u^n \cdot \nabla) \, u^n \cdot v \, dx - \int_\Omega f \cdot v \, dx, \ \forall v \in V_o ; \\ r_1^n \in V_o , \end{cases}$$

and set

$$(5.20) \quad w_1^n = r_1^n \, / \, (\alpha \int_\Omega | r_1^n |^2 dx + \nu \int_\Omega | \nabla r_1^n {}^2 | dx)^{1/2} .$$

Then, for $j = 2, \ldots, k$, *compute* r_j^n *and* w_j^n *via the solution of*

$$(5.21) \quad \begin{cases} \alpha \int_\Omega r_j^n \cdot v \, dx + \nu \int_\Omega \nabla r_j^n \cdot \nabla v \, dx = \langle DF(u_j^n ; w_{j-1}^n) , v \rangle \\ - \alpha \int_\Omega s_{j-1}^n \cdot v \, dx - \nu \int_\Omega \nabla s_{j-1}^n \cdot \nabla v \, dx, \ \forall v \in V_o ; \\ r_j^n \in V_o , \end{cases}$$

where

$$(5.22)\qquad \langle DF(u^n;w), v\rangle = \alpha\int_\Omega w\cdot v\,dx + \nu\int_\Omega \nabla w\cdot\nabla v\,dx + \int_\Omega (u^n\cdot\nabla)\,w\cdot v\,dx + \int_\Omega (w\cdot\nabla)\,u^n\cdot v\,dx,$$

$$(5.23)\qquad s^n_{j-1} = \sum_{i=1}^{j-1} b_{ij-1}\, w^n_i\,,$$

with

$$(5.24)\qquad b^n_{il} = \langle DF(u^n; w^n_l), w^n_i\rangle,$$

and by setting

$$(5.25)\qquad w^n_j = r^n_j \,/\, \Big(\alpha\int_\Omega |\, r^n_j\,|^2 dx + \nu\int_\Omega |\,\nabla r^n_j|^2 dx\Big)^{1/2}.$$

Step 2: Computation of u^{n+1}.

Solve in $\mathbf{R}^k$ *the following least squares problem*

$$(5.26)\qquad \begin{cases} \textit{Find } a^n = \{a^n_j\}_{j=1}^k \textit{ such that} \\ j_n(a^n) \le j_n(c),\ \forall c \in \mathbf{R}^k, \end{cases}$$

where

$$(5.27)\qquad j_n(c) = \frac{\alpha}{2}\int_\Omega |\, y\,|^2 dx + \frac{\nu}{2}\int_\Omega |\,\nabla y|^2\, dx,$$

with $(y = y^n(c))$ *the solution of the linear elliptic system*

$$(5.28)\qquad \begin{cases} \alpha y - \nu\nabla^2 y = \alpha u^n(c) - \nu\nabla^2 u^n(c) + (u^n(c)\cdot\nabla)u^n(c) - f \textit{ in } \Omega, \\ y = 0 \textit{ on } \Gamma, \end{cases}$$

$$(5.29) \qquad u^n(c) = u^n + \sum_{j=1}^{k} c_j\, w_j^n .$$

Finally

$$(5.30) \qquad u^{n+1} = u^n + \sum_{j=1}^{k} a_j\, w_j^n . \qquad \square$$

Do n = n+1 *and go to* (5.19) .

Remark 5.5: Problem (5.26) is conceptually very close to the nonlinear least squares formulations of the Navier-Stokes equations, discussed in [2], [3]; therefore, we can solve (5.26) by the conjugate gradient techniques described in the above references. However, in order to minimize the complexity of the above algorithms we shall take into account Remark 5.4 and use in the neighborhood of $c = 0$ a quadratic approximation of the functional in (5.26), (5.27). This simplification has been validated by numerical experiments ; actually, the performances of GMRES are improved if one uses a "back-tracking" strategy, like the one described in [1, Section 3] and in [13]. □

Generalizing the above methodology to compressible viscous flows is straightforward; it follows from Section 2 (and from [3]) that the system to be solved is given by

$$(5.31) \qquad \begin{cases} \alpha\sigma + u \cdot \nabla\sigma = g , \\ \alpha u - b\nabla^2 u - \psi(\sigma,u,T) = f , \\ \alpha T - b\Pi\, \Delta T - \chi(\sigma,u,T) = h , \end{cases}$$

with appropriate boundary conditions. In (5.31), α is a positive parameter and the functions g, f, h are given. Here, we shall use the GMRES algorithm of Section 5.2, taking for preconditioner the elliptic operator associated to the following norm

$$\{ \varphi, v, \theta\} \to \{\alpha \int_\Omega \varphi^2 dx + \alpha\, A \int_\Omega |v|^2 dx + b\mu\, A \int_\Omega |\nabla v|^2 dx$$
$$+ \alpha\, B \int_\Omega \theta^2 dx + b\Pi\, B \int_\Omega |\nabla\theta|^2 dx\}^{\frac{1}{2}} , \ A, B > 0,$$

(cf. [3] for more details) .

6. Comments on the compatibility between finite element approximations for velocity and density in the compressible case.

It is well known that in the case of the *incompressible* Navier-Stokes equations, *pressure* and *velocity* cannot be approximated independently (cf. [17] and the references therein) .

Typical choices are for example :

(i) *Continuous* and piecewise *linear* approximations for the *pressure*, associated to a *continuous* and piecewise *quadratic* approximation for the velocity.

(ii) *Continuous* and piecewise *linear* approximation for the *pressure*, associated to a *continuous* and piecewise *linear* approximation for the *velocity* on a grid *twice finer* than the pressure grid (see [2, Chapter 7] and [3, Section 5] for more details).

(iii) The so-called Arnold-Brezzi-Fortin *mini element* (cf. [18] and [3, Section 5]) .

Concerning now the *compressible* Navier-Stokes equations, there has been a natural tendency to approximate velocity and density (or logarithmic density) using the *same* finite element spaces for both quantities. Indeed, such a strategy can produce *spurious oscillations* particularly for the density. Once more, a most efficient cure is again to use approximations of type (i), (ii), (iii), with the role of the *pressure* played by the *density* (or the *logarithmic density*) ; as in the incompressible case, approximations (ii) and (iii) are particularly attractive, and therefore have been used for the subsequent calculations.

7. Numerical Experiments

7.1 Incompressible Viscous Flow.

We consider in this section the solution of the Stokes problem by the method discussed in section 4.1. The approximation of the Navier-Stokes equations is done by the finite element methods described in [2], [3]; we have used, therefore, approximations of the *pressure* which are *continuous* and *piecewise affine*, and similar approximations for the *velocity* over the *twice finer* grid obtained by joining the mid-points of the edges of the triangles of the pressure grid as shown on Figure 7.1 (see [2], [3] for more details).

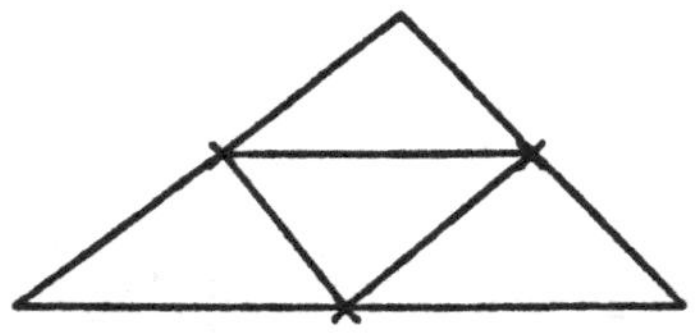

Figure 7.1

In view of checking the efficiency of algorithm (4.9) - (4.30), we have considered incompressible viscous flow around and inside the two-dimensional air intake of Figures 7.2 (a), 7.2 (b), 7.2 (c) where the pressure grids used for our calculations have been visualized. On Figures 7.3 (a) and 7.3 (b) we have shown for algorithm (4.9) - (4.30) the variation of

$$\varepsilon = 2 \ln \frac{\|g^n\|_{L^2(\Gamma)}}{\|g^0\|_{L^2(\Gamma)}},$$

versus the number n of iterations (Fig. 7.3 (a)), and also the C.P.U. time (fig. 7.3 (b)) for the three above space discretizations. We have also compared the performances of algorithm (4.9) - (4.30) (graphs 1′, 2′, 3′) with those of the standard algorithm (graphs 1, 2, 3) for which we have $S = I$ (this algorithm is described in e.g. [2], [4]). The above methodology is obviously applicable to the Navier-Stokes equations via the operator splitting methods discussed in section 3. Figures 7.4 (a) and 7.4 (b) show at $t = 1$ (for $1/\Delta t - \alpha = 10$) and for $Re = 1$ (graphs 1 and 1′) and $Re = 200$ (graphs 2 and 2′) the *convergence speed up*, measured in *iteration number* (Fig. 7.4 (a)) and *C.P.U. time* (Figure 7.4 (b)). We observe that the efficiency of preconditioning (4.6) increases with $Re/\Delta t$; this last property is quite interesting for flow at *high Reynolds number*, for which we have to take Δt sufficiently small to accurately simulate the evolution of unsteady phenomena.

7.2 Compressible Viscous Flows.

We discuss in this subsection, the results obtained for the numerical simulation of compressible viscous flows around and inside air intakes and after bodies.

We present first the performances of the iterative method ("compressible" variant of algorithm (4.9) - (4.30)) sketched in Section 4.2. We have been using here the following approximations:

(i) continuous and piecewise "P_1 + bubble" (in the sense of [3, Section 5]) for the *velocity* field.

(ii) Continuous and piecewise linear for *density* and *temperature*.

Figures 7.5 (a) and 7.5 (b) display the variation of the residual versus the number of iterations and the CPU time, respectively. The flow domain and the basic grids are those of the incompressible case discussed in the above Section 7.1. We observe a behaviour identical to the one of the incompressible case.

On Figures 7.6 (a) and 7.6 (b), we observe that, here too, both speeds of convergence increase with Re/Δt (in these numerical experiments the generalized Stokes solver is combined to nonlinear solvers via a time discretization of the compressible Navier-Stokes equations by a θ-scheme identical to the one in Section 3). On Figure 7.7 we have represented the variation of the cost (measured in CPU time) of the quasi-direct solver of the generalized Stokes problem (in the sense of [3, Section 12]) as a function of the number of boundary nodes; as we can see the function represented in Figure 7.7 increases fairly quickly.

The numerical experiments associated to compressible flows have also been a benchmark for comparisons between least squares/conjugate gradient methods (cf. [3]) and GMRES algorithms preconditioned by various linear operators (cf. Section 5.2). According to Figures 7.8 (a) and 7.8 (b) (conjugate gradient) and Figures 7.9 (a) and 7.9 (b) (GMRES), the GMRES algorithm appears to be significantly the faster one, both in iterations and CPU time (approximately by a factor of 3, here); these numerical experiments correspond to a compressible viscous flow around and inside the three dimensional afterbody of Figure 7.10.

On Figure 7.11, we have shown the convergence of GMRES, evaluated in CPU time, for various type of preconditioners and related solvers (REL: *Relaxation*, VDV: *incomplete Choleski factorization à la Van der Vorst*, DIAG: *diagonal preconditioner*).

We have also compared (see Figure 7.12) the performances of solvers for the generalized Stokes problem based on either preconditioning by a boundary operator (like in section 4.2 and the Appendix) or by a Neumann operator operating on the density (see [14] for the details). Figure 7.12 clearly shows an important "*starting up*" phenomenon for the first algorithm, which in first analysis seems to be related to the calculation of $\nabla \cdot \mathbf{f}$ in relation (4.50); on the other hand the first algorithm and the associated approximation method gives more accurate density and pressure than the second one.

In the case of the after body of Figure 7.10 we have shown on Figure 7.13 the distributions, in a meridian plane and at t = 9.5, of the Mach number, density, pressure and norm of the vorticity. The calculation has been done using a space approximation of $(P_1 + \text{bubble})/P_1$ type, a time discretization by a θ-scheme, and a GMRES algorithm for the nonlinearities.

The next test problem concerns a two dimensional compressible viscous flow around the elliptic body shown on Figure 7.14. We have taken here $M_\infty = .8$, $Re = 10^3$, $\gamma = 1.4$, $Pr = .72$ and a zero angle of attack. Figure 7.14 shows also a grid obtained by adaptive mesh refinement for this calculation (see, e.g., [3] for adaptive mesh refinement methods for the Navier-Stokes equations). We have shown on Figures 7.15 to 7.17 the isobar contours obtained with

(a) The same *continuous* and *piecewise linear* approximations for velocity and density (Figure 7.15).

(b) The approximation (ii) of Section 6 (Figure 7.16).

(c) The approximation (iii) of Section 6 (Figure 7.17).

We observe that the spurious oscillations associated to the first approximation disappear with the more sophisticated approximations (b) and (c). Similar phenomena are observed for three dimensional flow around spheres (see Figures 7.18, 7.19) and close to the forebody of the space vehicle Hermes at $M_\infty = 2$ and $Re = 10^3$ (see Figures 7.20, 7.21). For these three dimensional calculations mesh adaptive techniques have also been used. We observe that Figures 7.18 and 7.19 correspond to simulations close to a stagnation point where very strong compression phenomena take place.

8. Conclusion

It follows from the above sections that efficient solutions methods for the simulation of viscous flow are obtained via the combination of

(i) Time discretization by θ-schemes,

(ii) Finite element methods for the space approximation using different representation for the velocity and the pressure (or the density),

(iii) GMRES algorithms to treat the nonlinearities,

(iv) Preconditioned conjugate gradient algorithms for the generalized Stokes problems.

We strongly believe that substantial convergence accelerations will be obtained through a systematic use of multigrid algorithms. The current and future applications concern civil and military aircrafts and also some parts of the flights of space vehicles.

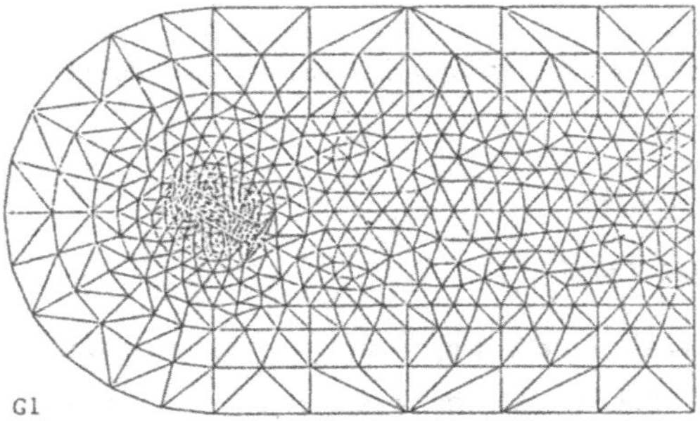

(a) 182 boundary nodes

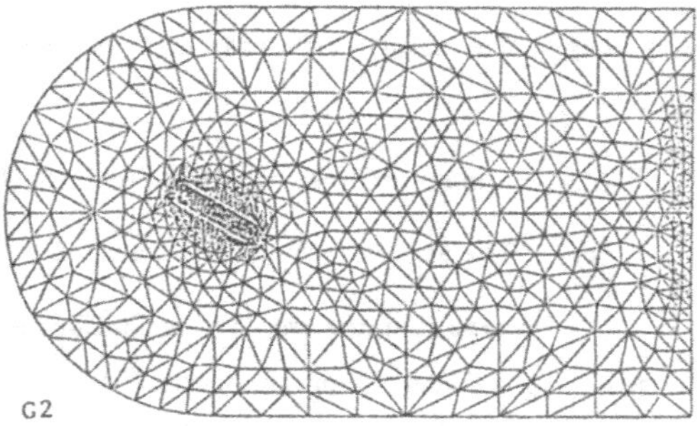

(b) 204 boundary nodes

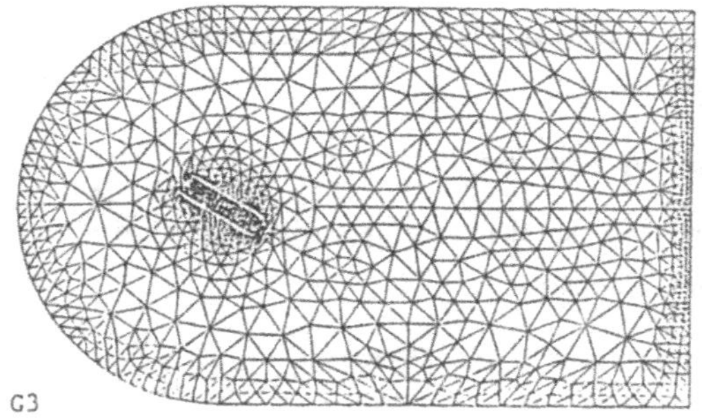

(c) 276 boundary nodes

Figure 7.2.

Pressure grids.

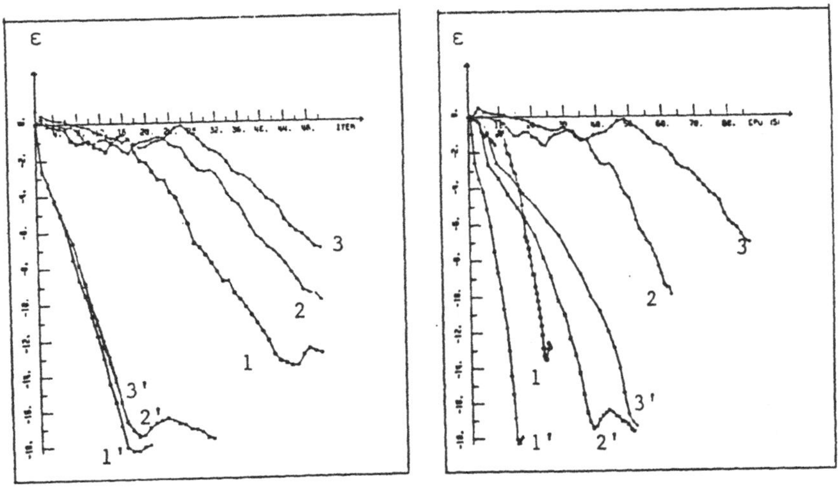

(a) (b)

Figure 7.3.

Influence of the grid. 1,1' on G1 ;
2,2' on G2 ; 3,3' on G3

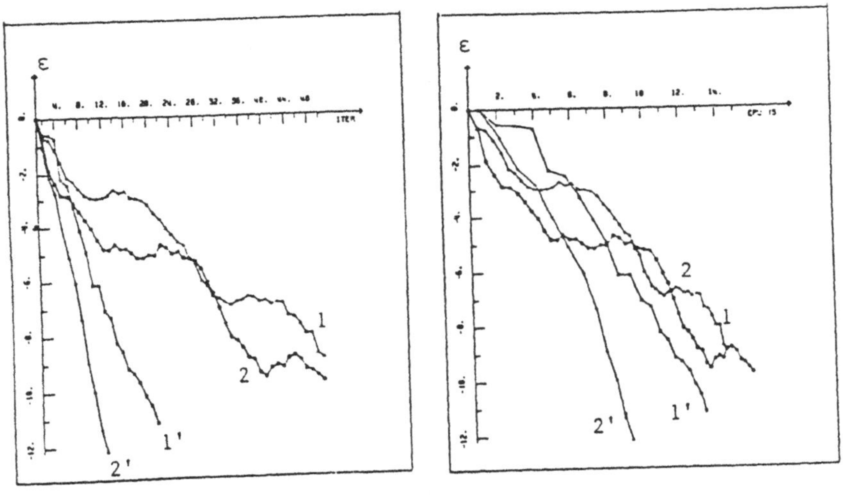

(a) (b)

Figure 7.4.

Influence of the Reynolds number on the efficiency of
the preconditioning Re = 1. for 1,1' ; Re = 200 for 2,2'.

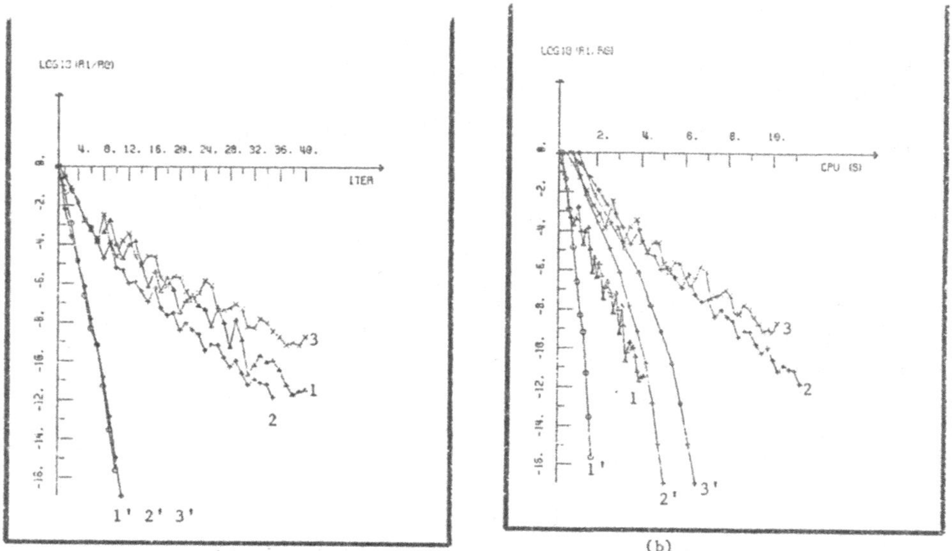

(a) (b)

Figure 7.5.

Compressible Stokes problem

Influence of the grid. 1,1' on G1 ; 2,2' on G2 ; 3,3' on G3.

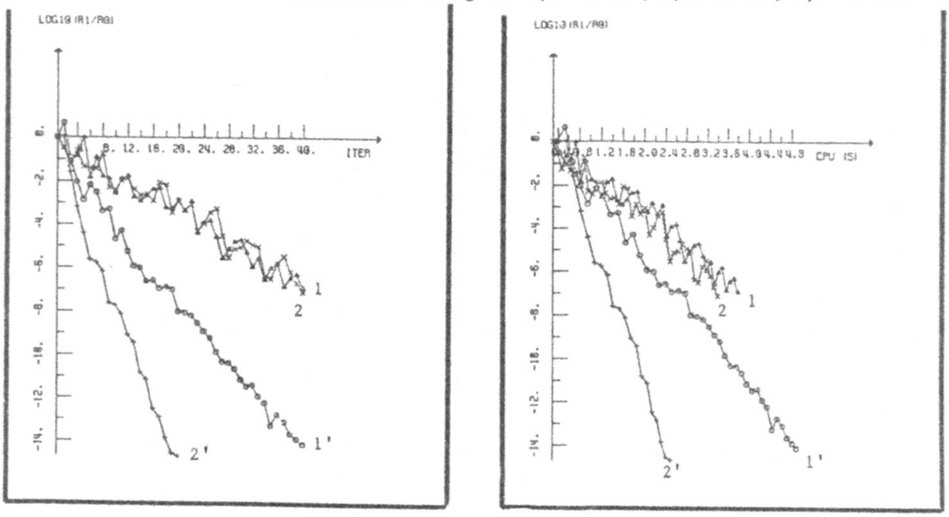

(a) (b)

Figure 7.6.

Influence of the Reynolds number on the efficiency of the preconditioning. Re = 500 for 1,1' ; Re = 10000. for 2,2'.

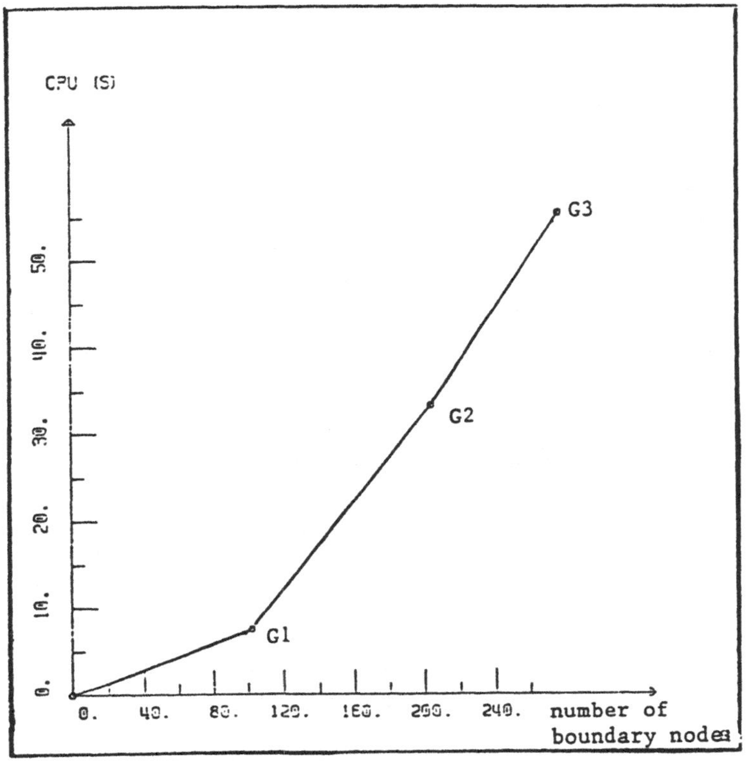

Figure 7.7.

Compressible Stokes problem. Quasi direct solution. Influence of the number of boundary nodes on the computational cost of the boundary operator.

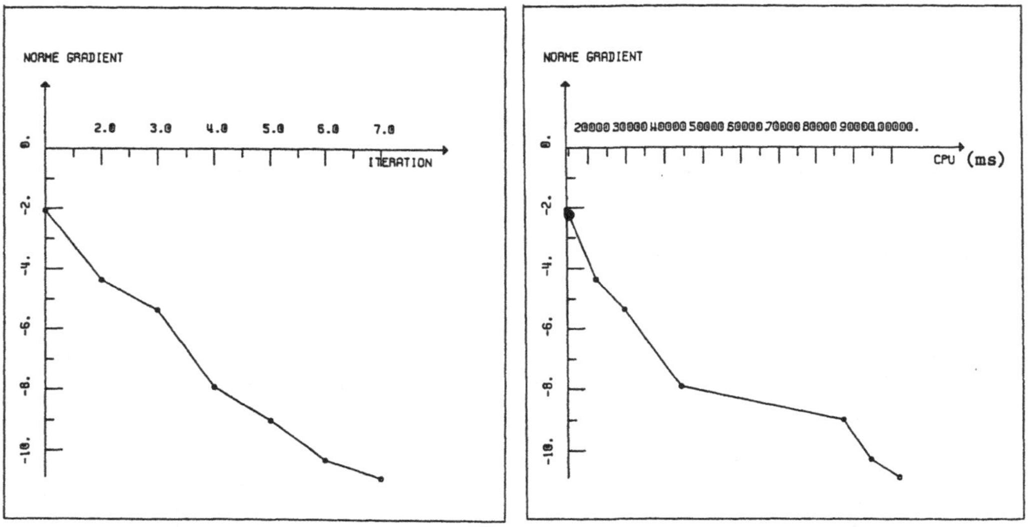

(a) (b)

Figure 7.8.

Least squares/conjugate gradient algorithm.

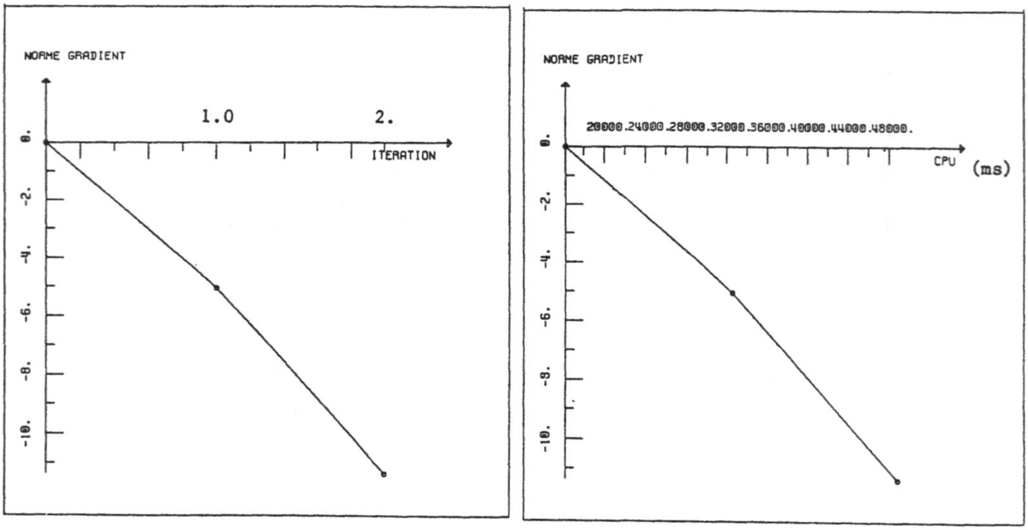

(a) (b)

Figure 7.9.

GMRES algorithm.

Figure 7.10.
3D Mesh around
and inside an afterbody

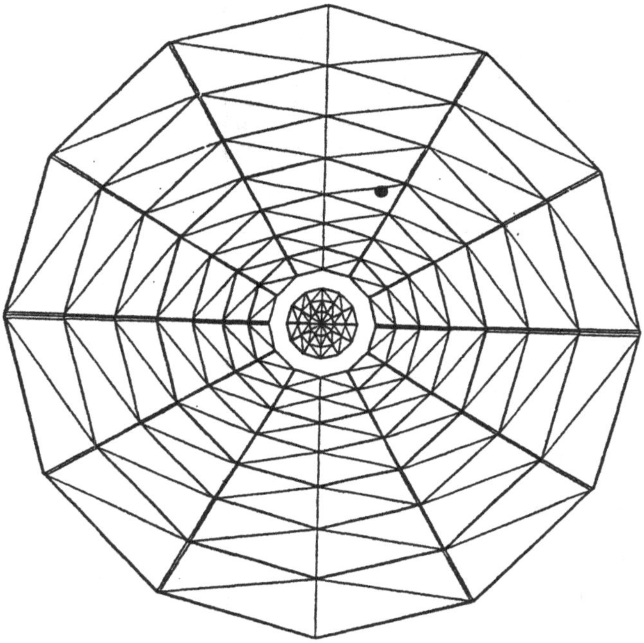

a) Cross section

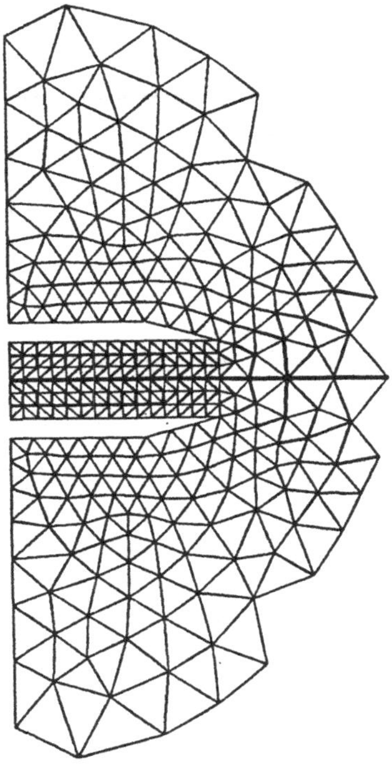

b) Meridian section

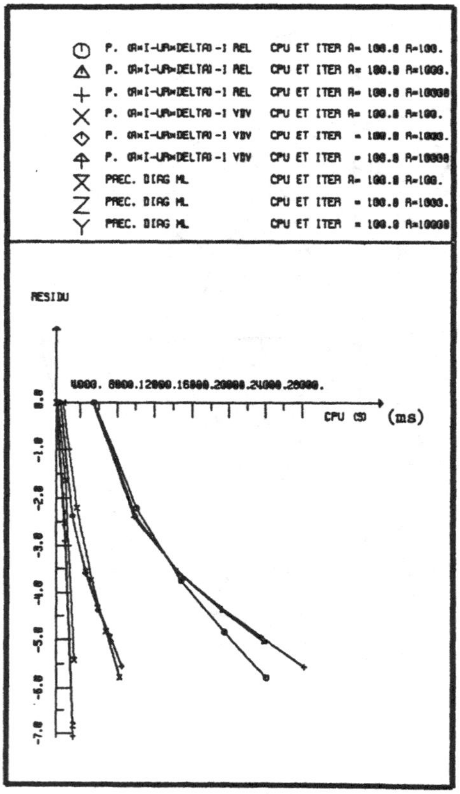

Figure 7.11.
Convergence of GMRES with different preconditionners and related solvers.

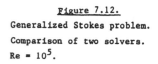

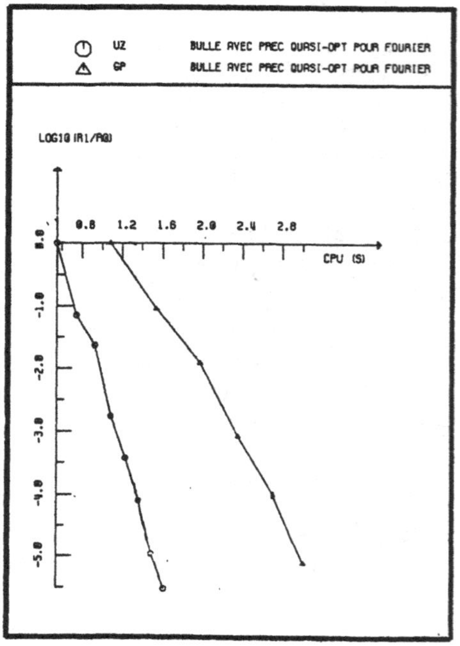

Figure 7.12.
Generalized Stokes problem.
Comparison of two solvers.
Re = 10^5.

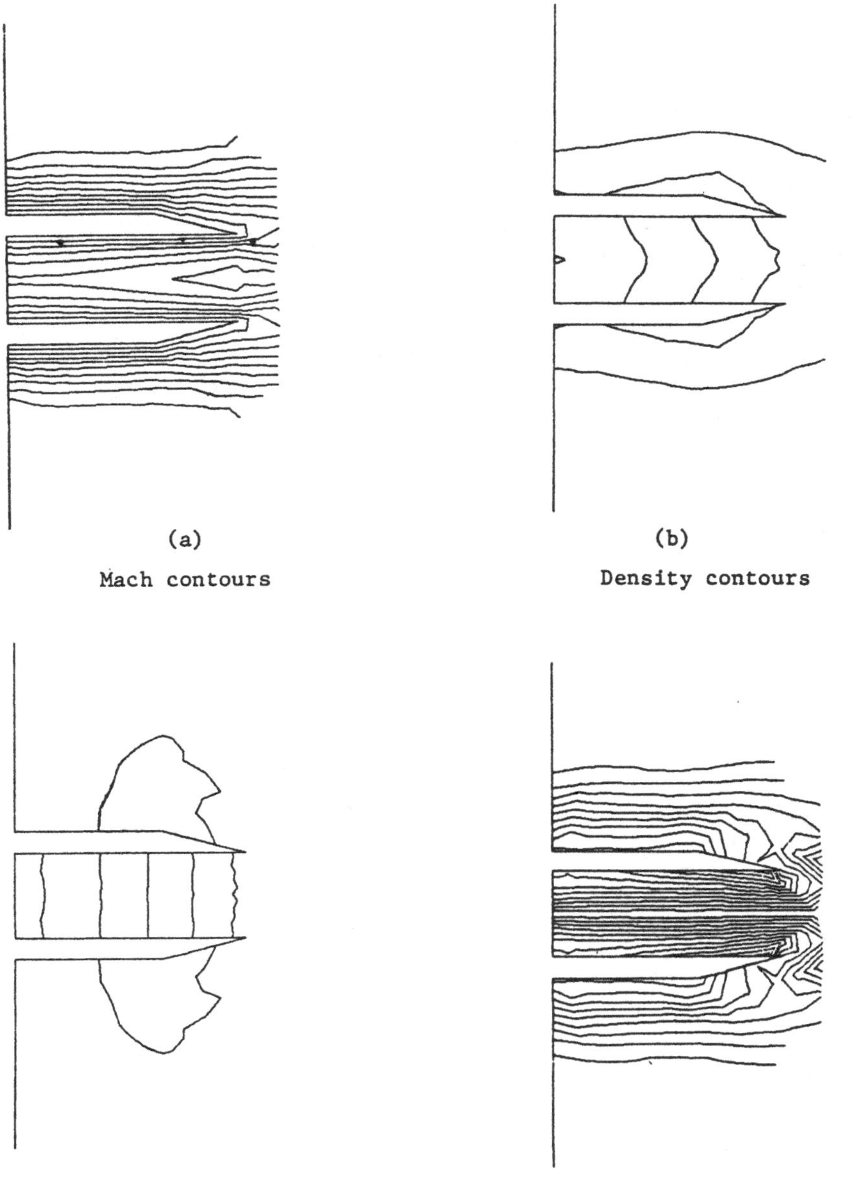

(a) Mach contours

(b) Density contours

(c) Pressure contours

(d) Vorticity contours

Figure 7.13.

3D computation around and inside an afterbody.

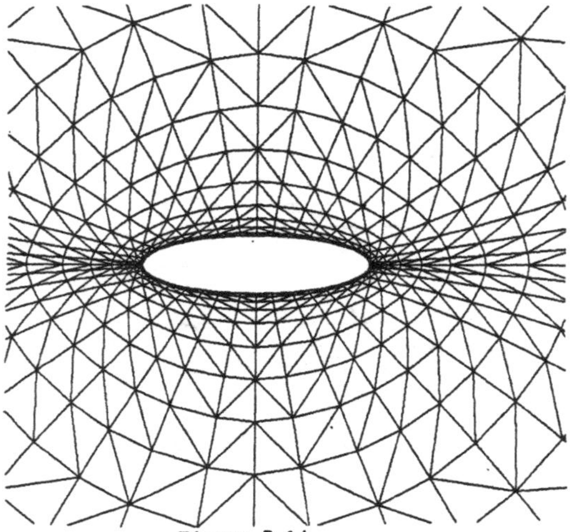

Figure 7.14.
Adapted grid around an elliptic body.

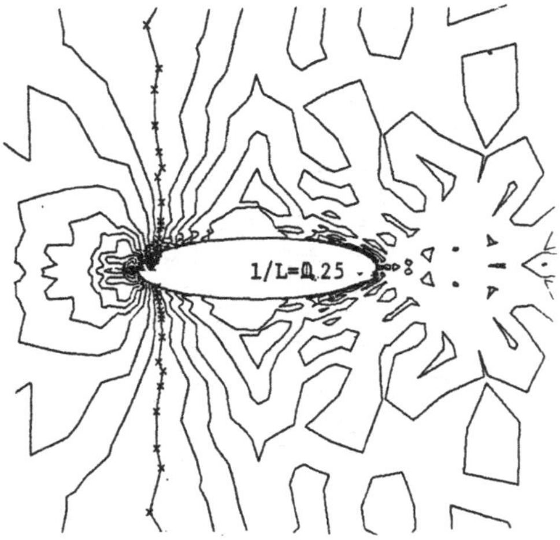

Figure 7.15.
P_1 approximations.

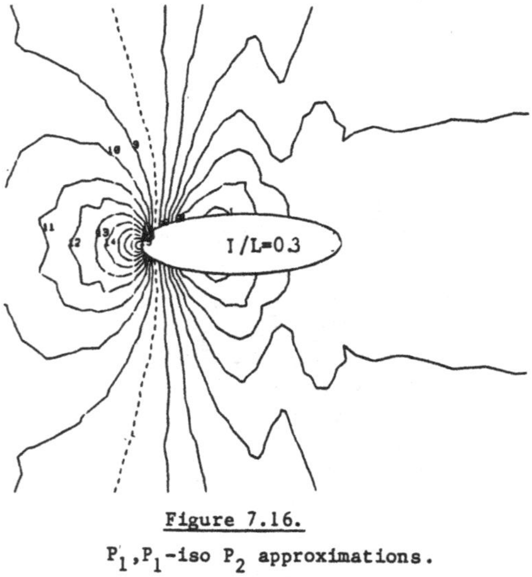

Figure 7.16.
P_1, P_1-iso P_2 approximations.

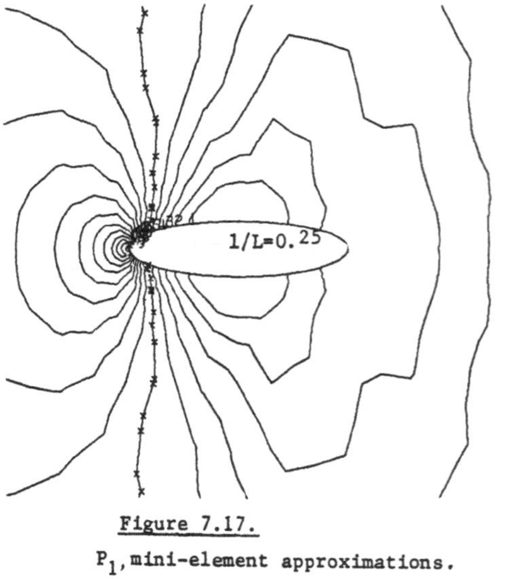

Figure 7.17.
P_1, mini-element approximations.

Flow around an elliptic body
Isobar contours. $M_\infty = 0.8$, Re $= 10^3$.

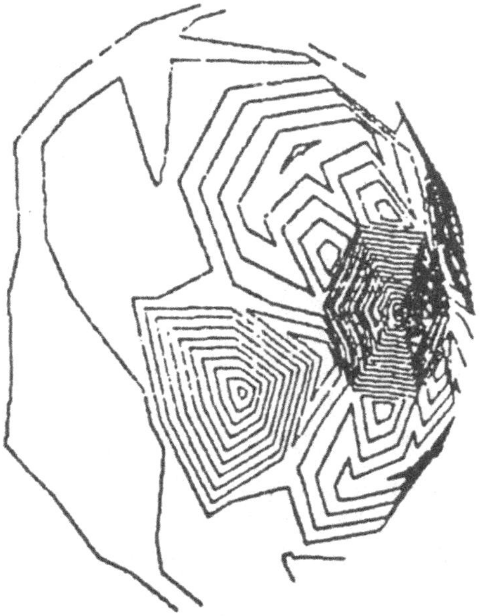

Figure 7.18.
P_1 approximations

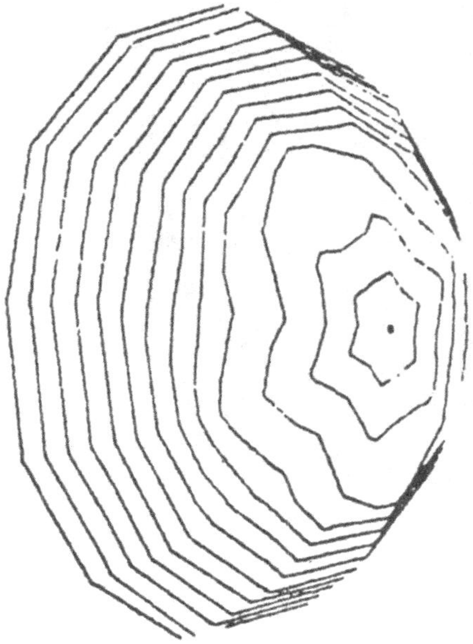

Figure 7.19.
P_1, mini-element
approximations.
Flow around a sphere
$M_\infty = 0.9$, $Re = 10^2$.

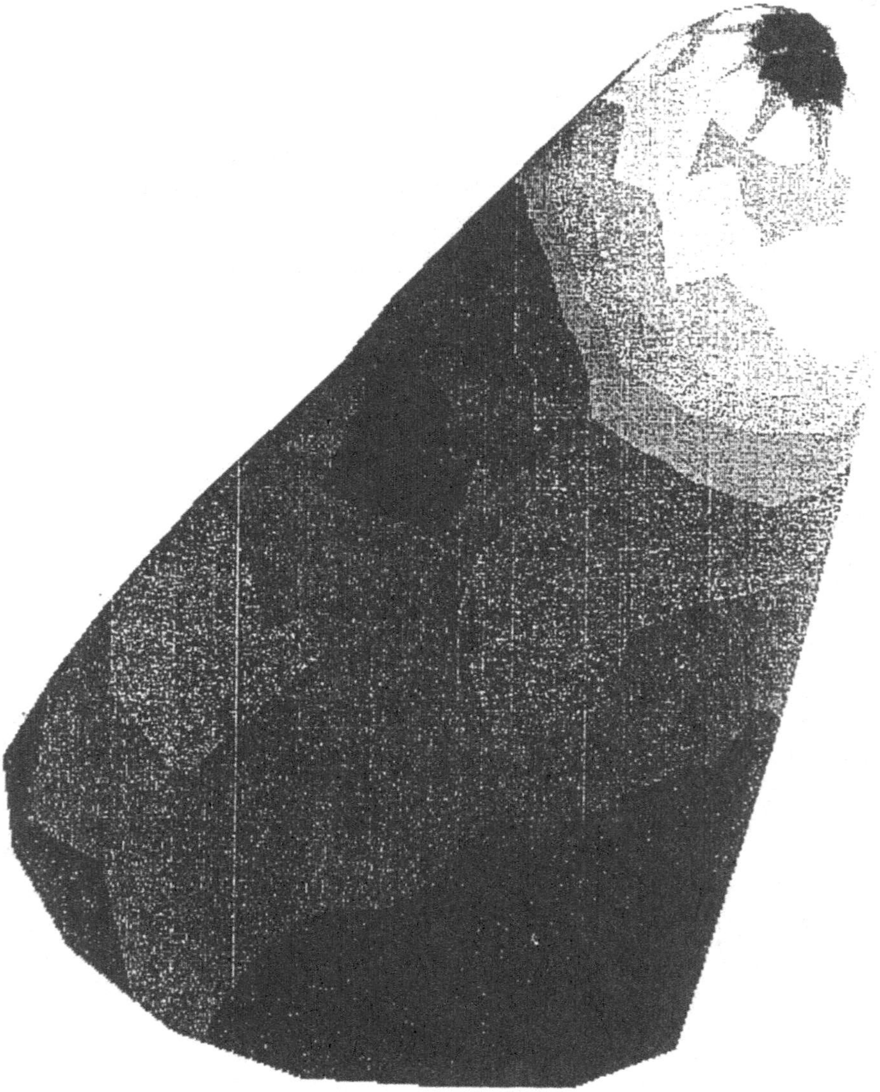

Figure 7.20.

P_1 approximations.

Flow around the forebody of Hermès. $M_\infty = 2$, $Re = 10^3$.

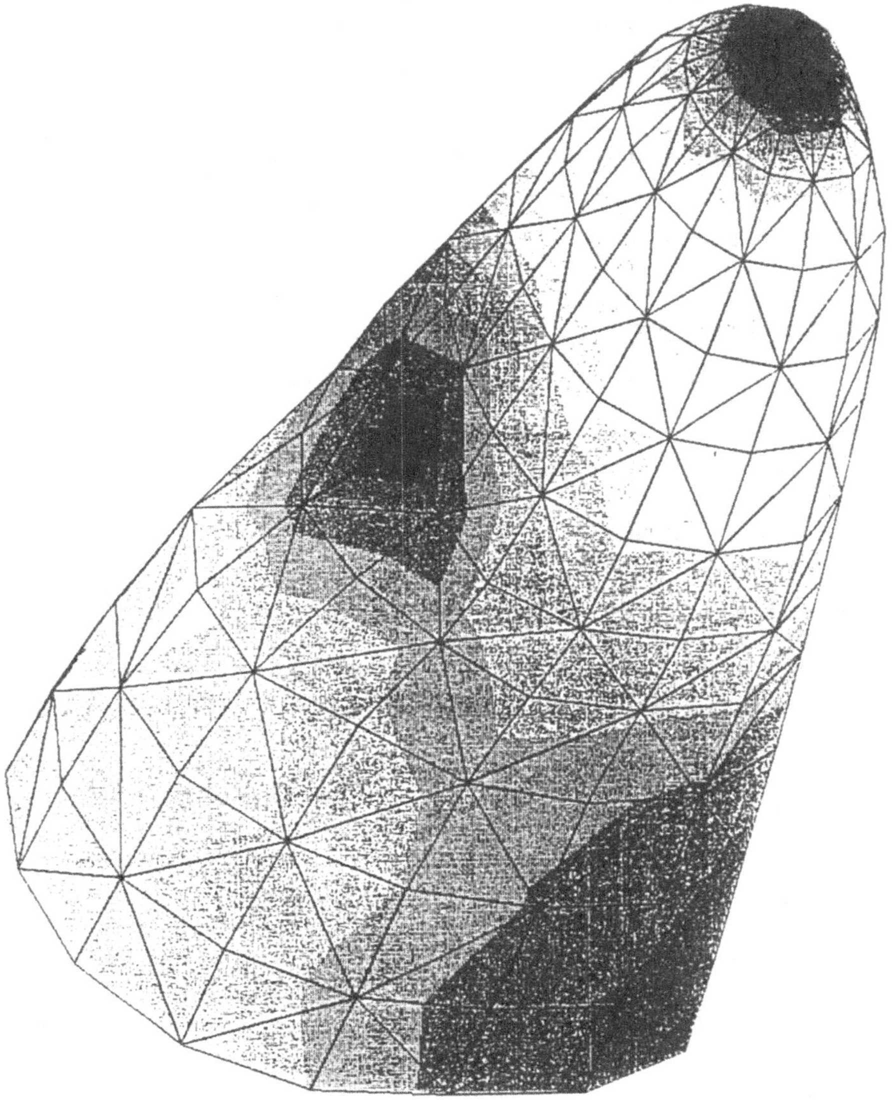

Figure 7.21.

P_1, mini-element approximations.

Appendix

Application of Fourier Analysis to the construction of a preconditioning operator for the generalized Stokes problems originating from compressible viscous flow calculations.

1. Motivation.

The goal of this appendix is to justify the various formulas given in Section 4.2 for the generalized Stokes problems encountered in the numerical treatment of compressible viscous flows (for the incompressible case, cf. [5], [6]).

2. The generalized Stokes problem.

It follows from Section 4.2 that the generalized Stokes problem under consideration is defined by

(2.1) $$\alpha\sigma + \nabla\cdot\mathbf{u} = g \text{ in } \Omega ,$$

(2.2) $$\alpha\mathbf{u} - a\mu\nabla^2\mathbf{u} + \beta\nabla\sigma = \mathbf{f} \text{ in } \Omega ,$$

(2.3) $$\mathbf{u} = 0 \text{ on } \Gamma_B, \quad \mathbf{u} = \mathbf{u}_\infty \text{ on } \Gamma_\infty^-, \quad a\mu\frac{\partial\mathbf{u}}{\partial\mathbf{n}} - \beta\mathbf{n}\sigma = \mathbf{s} \text{ on } \Gamma_\infty^+.$$

In this appendix, we shall consider Dirichlet boundary conditions only; it can then be shown (cf. Section 4.2) that the trace λ of σ on Γ is the solution of a problem associated to the *boundary operator* A defined as follows:

(2.4) $$\begin{cases} \alpha^2\sigma_\lambda - \zeta\nabla^2\sigma_\lambda = 0 \text{ in } \Omega , \\ \sigma_\lambda = \lambda \text{ on } \Gamma , \end{cases}$$

(2.5) $$\begin{cases} \alpha\mathbf{u}_\lambda - a\mu\nabla^2\mathbf{u}_\lambda = -\beta\nabla\sigma_\lambda \text{ in } \Omega , \\ \mathbf{u}_\lambda = 0 \text{ on } \Gamma , \end{cases}$$

(2.6)
$$\begin{cases} \alpha^2\psi_\lambda - \zeta\nabla^2\psi_\lambda = \alpha\sigma_\lambda + \nabla\cdot \mathbf{u}_\lambda \ \textit{in}\ \Omega\,, \\ \psi_\lambda = 0 \quad \textit{on}\ \Gamma\,, \end{cases}$$

(2.7)
$$A\lambda = -\frac{\partial\psi_\lambda}{\partial n}\Big|_\Gamma$$

with

(2.8)
$$\zeta = \beta + \alpha a\mu\,.$$

3. Determination of A when Ω is an half-space.

Operator A is a *convolution* operator; to identify its *symbol* we shall compute $Ae^{2i\Pi sx_2}$ in the particular case where

(3.1)
$$\Omega = \{x|x = \{x_1, x_2\}\ \varepsilon\ \mathbf{R}^2, x_1 > 0,\ x_2 \varepsilon \mathbf{R}\}\,.$$

We first have to solve

(3.2)
$$\begin{cases} \alpha^2\sigma - \zeta\nabla^2\sigma = 0 \ \textit{in}\ \Omega\,, \\ \sigma = e^{2i\Pi sx_2} \ \textit{on}\ \Gamma\,, \end{cases}$$

whose solution (obtained by a *separation of variables* technique) is given by

(3.3)
$$\sigma = e^{-\omega x_1} e^{2i\Pi sx_2}\,,$$

where

(3.4)
$$\omega^2 = \frac{\alpha^2}{\zeta} + 4\Pi^2 s^2\,.$$

The solution of (2.5) with σ defined by (3.3), (3.4) yields

$(3.5)_1 \qquad u = \{u_1, u_2\}$,

with

$(3.5)_2 \qquad u_1 = \frac{\omega\zeta}{\alpha}\left(e^{-\omega x_1} - e^{-\chi x_1}\right) e^{2i\Pi s x_2}$,

$(3.5)_3 \qquad u_2 = -\, 2i\Pi s \frac{\zeta}{\alpha}\left(e^{-\omega x_1} - e^{-\chi x_1}\right) e^{2i\Pi s x_2}$,

with

$(3.6) \qquad \chi^2 = \frac{\alpha}{a\mu} + 4\Pi^2 s^2$.

Next, we solve (2.6) with σ and u given by (3.3), (3.5); we obtain then

$(3.7) \qquad \psi = -\frac{a\mu\zeta}{\alpha^2\beta}\left(\omega\chi - 4\Pi^2 s^2\right)\left(e^{-\chi x_1} - e^{-\omega x_1}\right) e^{2i\Pi s x_2}$.

We have then

$(3.8) \qquad A\, e^{2i\Pi s x_2} = -\frac{\partial\psi}{\partial n}\Big|_\Gamma = \frac{\partial\psi}{\partial x_1}\Big|_{x_1=0} = \dfrac{\frac{\alpha^2}{a\zeta\mu} + 4\Pi^2 s^2\left(\frac{1}{a\mu} + \frac{\alpha}{\zeta}\right)}{(\omega+\chi)\,(\omega\chi + 4\Pi^2 s^2)}\, e^{2i\Pi s x_2}$.

The symbol of A is therefore given by

$(3.9) \qquad \mathcal{A}(s) = \dfrac{\frac{\alpha^2}{a\zeta\mu} + 4\Pi^2 s^2\left(\frac{1}{a\mu} + \frac{\alpha}{\zeta}\right)}{(\omega+\chi)\,(\omega\chi + 4\Pi^2 s^2)}$.

We look now for $S = B^{-1}$, such that the symbol $\mathcal{B}(s)$ of B verifies

$(3.10) \qquad d\,\frac{(\omega+c)^2}{\omega} = \mathcal{B}(s)$.

Then we adjust d and c in such a way that

(3.11) $\lim_{s \to 0} \mathcal{A}(s)\, \mathcal{B}(s) = \lim_{|s| \to +\infty} \mathcal{A}(s)\, \mathcal{B}(s)$.

From (3.9) - (3.11) we obtain

(3.12) $d = 4/\left(\frac{\alpha}{\zeta} + \frac{1}{a\mu}\right)$

and

(3.13) $c = \frac{\alpha}{\sqrt{\zeta}} \left\{ \frac{1}{2}\left[\left(1 + \frac{\zeta}{\alpha a \mu}\right)\left(1 + \sqrt{\frac{\alpha a \mu}{\zeta}}\right)\right]^{\frac{1}{2}} - 1 \right\}$.

Operator $S = B^{-1}$ will be a good preconditioner if

(3.14) $\mathcal{A}(s)\, \mathcal{B}(s)$ *is close to* 1, $\forall s \in \mathbf{R}$.

Figures A1 and A2 show, for typical values of the parameters, the variation of $\mathcal{A}(s)\, \mathcal{B}(s)$ as a function of $|s|$. From these graphs, we can hope a good speed convergence, since the conjugate gradient algorithm preconditioned by $S = B^{-1}$ will have as contraction coefficient

(3.15) $\text{Coef} = \frac{\sqrt{\nu_{AB}} - 1}{\sqrt{\nu_{AB}} + 1}$,

where ν_{AB} is the maximal value in the above graphs. If, for example, $\nu_{AB} = 2$ (resp. 1.5) the above coefficient is 1/6 (resp. 0.101 . . .), which is remarkable.

The effective realization of B is described in Section 4.2 and is given by relations (4.54) - (4.58).

Acknowledgment: The authors would like to thank C. Begue, B. Mantel and G. Rogé for their invaluable help in the preparation of the present document. They have also taken advantage of several discussions with Drs. F. Brezzi, M. Fortin and O. Pironneau. The partial support of DRET (Grant #88.103) and CCVR (since part of the computations have been done on the CRAY2 of CCVR) is also acknowledged. Last, but not least, the authors would like to thank Sherry Nassar for her beautiful processing of the manuscript.

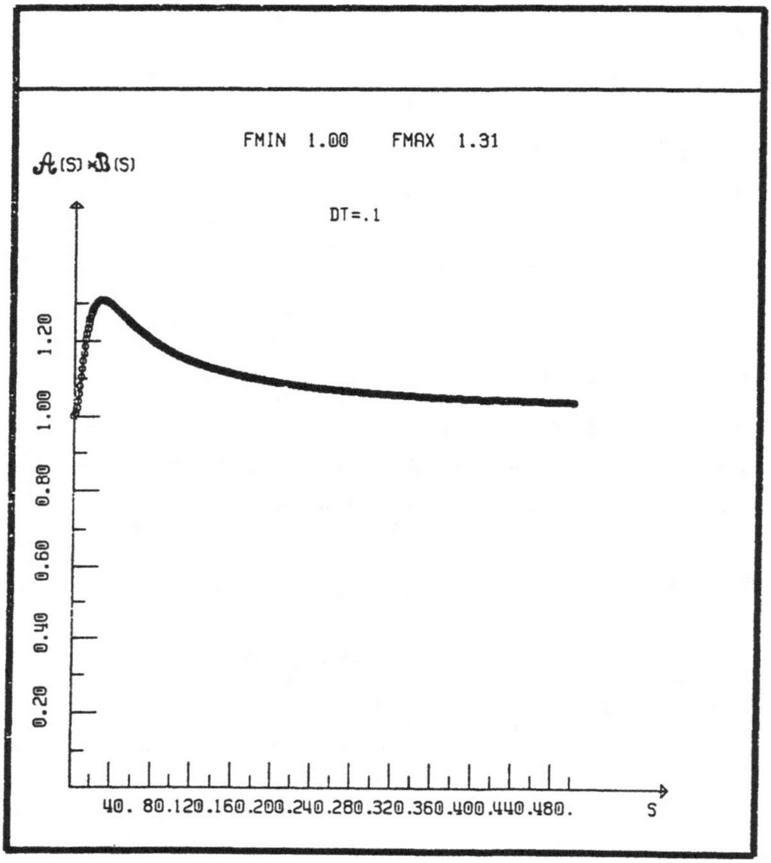

Figure A.1.

(M_∞ = .9, Re = 500)

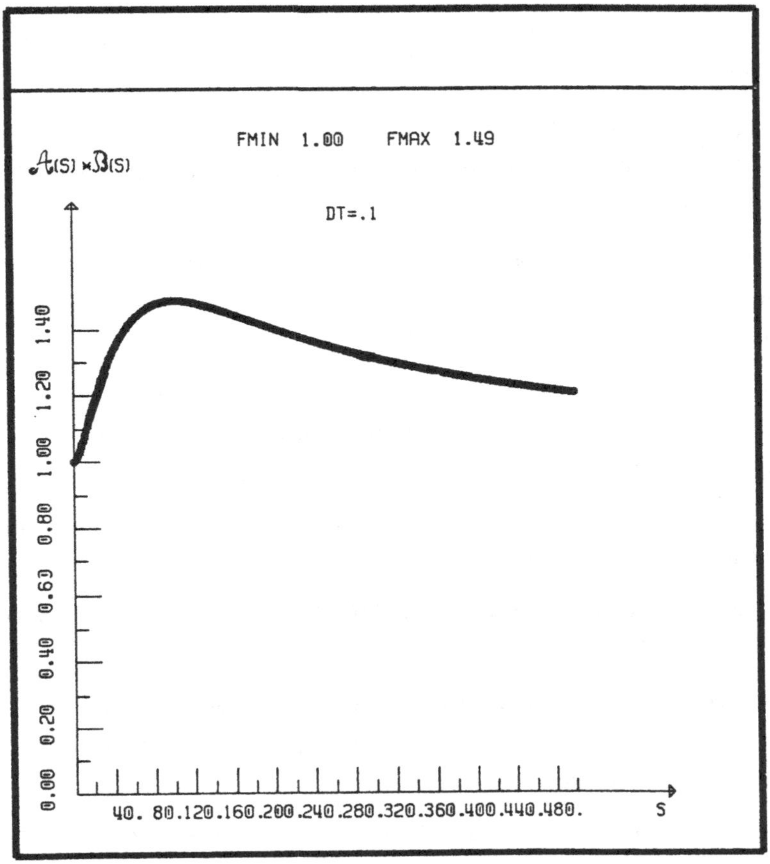

Figure A.2.
(M_∞ = .9, Re = 10000)

References

[1] P. N. Brown, Y. Saad, Hybrid Krylov Methods for Nonlinear Systems of equations, *Lawrence Livermore National Laboratory Research Report UCLR*-97645, Nov. 1987.

[2] R. Glowinski, *Numerical Methods for Nonlinear Variational Problems*, Springer-Verlag, New York, 1984.

[3] M. O. Bristeau, R. Glowinski, J. Périaux, Numerical Methods for the Navier-Stokes Equations. Application to the Simulation of Compressible and Incompressible Viscous Flows, *Computer Physics Reports*, 6, (1987), pp. 73-187.

[4] R. Glowinski, O. Pironneau, On numerical methods for the Stokes problem, Chapter 13 of *Energy Methods in Finite Element Analysis*, R. Glowinski, E. Y. Rodin, O. C. Zienkiewicz eds., J. Wiley, Chichester, 1979, pp. 243-264.

[5] C. Bègue, R. Glowinski, J. Périaux, Détermination d'un opérateur de préconditionnement pour la résolution itérative du problème de Stokes dans la formulation d'Helmholtz, *C. R. Acad. Sciences*, Paris, T.306, Série I, pp. 247-252, 1988.

[6] R. Glowinski, On a new preconditioner for the Stokes problem, *MatematicaAplicada e Computational*, 6, (1987), 2, pp. 123-140.

[7] N. Goutal, *Résolution des équations de Saint-Venant*, Ph.D. dissertation, University of Paris VI, Paris, France, Feb. 1987.

[8] C. Bègue, M. O. Bristeau, R. G. Glowinski, B. Mantel, J. Périaux, G. Rogé, Sur l'Analyse de Fourier d'un opérateur de préconditionnement pour le problème de Stokes généralisé des écoulements visqueux compressibles (to appear).

[9] M. O. Bristeau, O. Pironneau, R. Glowinski, J. Périaux, P. Perrier and G. Poirier, On the numerical solution of nonlinear problems in Fluid Dynamics by least squares and finite element methods (II). Application to transonic flows simulations, *Comp. Meth. in Appl. Mech. Eng.*, 51, (1985), pp. 363-394.

[10] L. B. Wigton, N. J. Yu and D. P. Young, GMRES Acceleration of Computational Fluid Dynamics Codes, *AIAA 7th Computational Fluid Dynamics Conference*, Cincinnati, Ohio July, 1985, Paper 85-1494, pp. 67-74.

[11] C. Bègue, M. O. Bristeau, R. Glowinski, B. Mantel, J. Périaux, Acceleration of the convergence for viscous flow calculations, in *Numeta 87, Vol, 2*, C. N. Pande, J. Middleton eds., Martinus Nighoff Publishers, Dordrecht, 1987, pp. T4/1 - T4/20.

[12] M. Mallet, J. Périaux, B. Stoufflet, On fast Euler and Navier-Stokes solvers, *Proceedings of the 7th GAMM Conference on Numerical Methods in Fluid Mechanics*, Louvain, Belgium, 1987.

[13] J. E. Dennis, R. B. Schnabel, *Numerical Methods for Unconstrained Optimization and Nonlinear Equations*, Prentice Hall, Englewood Cliffs, N. J. 1983.

[14] C. Rogé, Ph.D. dissertation, University of Paris VI, Paris, France, 1988/1989 (to appear).

[15] M. O. Bristeau, R. Glowinski, B. Mantel, J. Périaux, C. Rogé, Self- adaptive finite element method for 3D compressible Navier-Stokes flow simulation in Aerospace Engineering, *Proceedings of the 11th Int. Conf. on Num. Meth. in Fluid Dynamics*, Williamsburg, USA, June 1988 (to appear).

[16] J. Cahouet, J. P. Chabard, Multi-domains and multi-solvers finite element approach for the Stokes problem, in *Innovative Numerical Methods in Engineering*, R. P. Shaw et al, eds., Springer-Verlag, Berlin, 1986, pp. 317-322.

[17] V. Girault, P. A. Raviart, *Finite Element Methods for Navier-Stokes Equations*, Springer-Verlag, Berlin, 1986.

[18] D. N. Arnold, F. Brezzi, M. Fortin, A stable finite element for the Stokes equations, *Calcolo*, 21, (1984), 337.

[19] M. O. Bristeau, R. Glowinski, B. Mantel, J. Périaux, G. Roge, Acceleration of compressible Navier-Stokes calculations, to appear in the *Proceedings of the IMA International Conference on Computational Fluid Dynamics*, Oxford, 1988.